AF903418

BIOLUMINESCENCE – RECENT ADVANCES IN OCEANIC MEASUREMENTS AND LABORATORY APPLICATIONS

Edited by **David Lapota**

Bioluminescence – Recent Advances in Oceanic Measurements and Laboratory Applications
Edited by David Lapota

Published by InTech
Janeza Trdine 9, 51000 Rijeka, Croatia

Copyright © 2011 InTech
All chapters are distributed under the Creative Commons Attribution 3.0 license, which allows users to download, copy and build upon published articles even for commercial purposes, as long as the author and publisher are properly credited, which ensures maximum dissemination and a wider impact of our publications. After this work has been published by InTech, authors have the right to republish it, in whole or part, in any publication of which they are the author, and to make other personal use of the work. Any republication, referencing or personal use of the work must explicitly identify the original source.

As for readers, this license allows users to download, copy and build upon published chapters even for commercial purposes, as long as the author and publisher are properly credited, which ensures maximum dissemination and a wider impact of our publications.

Notice
Statements and opinions expressed in the chapters are these of the individual contributors and not necessarily those of the editors or publisher. No responsibility is accepted for the accuracy of information contained in the published chapters. The publisher assumes no responsibility for any damage or injury to persons or property arising out of the use of any materials, instructions, methods or ideas contained in the book.

Publishing Process Manager Martina Durovic

Technical Editor Teodora Smiljanic

Cover Designer InTech Design Team

Additional hard copies can be obtained from orders@intechopen.com

Bioluminescence – Recent Advances in Oceanic Measurements and Laboratory Applications, Edited by David Lapota
p. cm.
ISBN 978-953-307-940-0

We are IntechOpen,
the world's leading publisher of
Open Access books
Built by scientists, for scientists

3,900+
Open access books available

116,000+
International authors and editors

120M+
Downloads

151
Countries delivered to

Our authors are among the
Top 1%
most cited scientists

12.2%
Contributors from top 500 universities

WEB OF SCIENCE™

Selection of our books indexed in the Book Citation Index
in Web of Science™ Core Collection (BKCI)

Interested in publishing with us?
Contact book.department@intechopen.com

Numbers displayed above are based on latest data collected.
For more information visit www.intechopen.com

For EU product safety concerns:
IN TECH d.o.o., Prolaz Marije Krucifikse Kozulić 3, 51000 Rijeka, Croatia,
info@intechopen.com or visit our website at intechopen.com.

Meet the editor

Dr Lapota is the US Navy's leading scientist in bioluminescence research activities and developer of bioluminescence measurement systems. He has worked at the Space and Naval Warfare Systems Center, Pacific for 33 years. His academic credentials include a PhD in Biology from University of California, Santa Barbara, an MA in Geography and a BS in Zoology from San Diego State University. His accomplishments include more than 30 published manuscripts in the open literature, 37 symposia presentations, 22 oceanographic expeditions and 10 US Navy patents in bioluminescence instrumentation. Dr Lapota was also awarded two Exemplary Achievement Awards, and the Outstanding Technology Transfer Professional Award from the Federal Laboratory Consortium, for the Qwiklite 200, from Department of Defense in 2008, at SSC PACIFIC. Dr Lapota is a member of the American Geophysical Union (Ocean Sciences Member, 1981-present) Washington, D.C., American Society of Testing and Materials (Committee E47 - Biological Effects), The Explorers Club, Fellow National (Elected 1989), and the Marine Technology Society.

Contents

Preface

As someone who has spent more than 33 years studying the bioluminescence phenomenon in the world's oceans, I am continuously amazed by the many bioluminescence adaptations marine and terrestrial animals have developed to ensure their existence. It can hardly be considered a random occurrence as it has developed among various types of organisms, such as single celled dinoflagellates to the much more complex forms such as shrimp, fish, squid beetles, and worms. Bioluminescence has many functions, from predator-prey interactions and courtship, to camouflage and alert status from potential predators.

We now find ourselves utilizing luciferase – luciferin proteins, ATP, genes and the whole complexities of these interactions to observe and follow the progress or inhibition of tumors in animal models by measuring bioluminescence intensity, spatially and temporally using highly sophisticated camera systems. The following chapters describe applications in preclinical oncology research by bioluminescence imaging (BLI) with a variety of applications. Two other chapters describe current methodologies for rapid detection of contaminants using the Milliflex system, and the use of bioluminescence resonance energy transfer (BRET) technology for monitoring physical interactions between proteins in living cells. Others are using bioluminescent proteins for high sensitive optical reporters imaging in living animals, developing pH-tolerant luciferase for brighter *in vivo* imaging, and oscillation characteristics in bacterial bioluminescence. Lastly, using recent data, two chapters describe the long-term seasonal characteristics of oceanic bioluminescence and the responsible planktonic species producing bioluminescence. Such studies are few and rare.

I hope that after you read these chapters, many more questions will come to mind, which will encourage further studies into this fascinating area.

Dr David Lapota
Space and Naval Warfare Systems Center, Pacific
San Diego, California
U.S.A.

Part 1

Oceanic Measurements of Bioluminescence

1

Long Term Dinoflagellate Bioluminescence, Chlorophyll, and Their Environmental Correlates in Southern California Coastal Waters

David Lapota
Space and Naval Warfare Systems Center, Pacific
USA

1. Introduction

While many oceanographic studies have focused on the distribution of bioluminescence in the marine environment (Stukalin 1934, Tarasov 1956, Seliger et al. 1961, Clarke and Kelly 1965, Bityukov 1967, Lapota and Losee 1984, Swift et al. 1985, Lapota et al. 1988, Batchelder and Swift 1989, Lapota et al. 1989, Lapota and Rosenberger 1990, Neilson et al. 1995, Ondercin et al. 1995, Swift et al. 1995), little understanding of the seasonality and sources of planktonic bioluminescence in coastal waters and open ocean has emerged. Some previous studies with respect to annual cycles of bioluminescence were severely limited in duration as well as in the methods used to quantify bioluminescence (Bityukov 1967, Tett 1971). Only a few studies have measured bioluminescence on an extended basis, and these were short in duration, usually less than 2 years with long intervals between sets of measurements (Bityukov 1967, Yentsch and Laird 1968, Tett 1971). Others report data collected at different times of the year (Batchelder and Swift 1989, Batchelder et al. 1992, Buskey 1991) but do not address the seasonality of bioluminescence. Thus the detailed temporal variability of bioluminescence has never been characterized continuously over several years. Lack of such long-term studies leaves unanswered important questions regarding the role of bioluminescence in successional phenomena.

To adequately understand, model, and predict planktonic bioluminescence in any ocean, measurements must be conducted on a continual basis for at least several years in order to evaluate intra- and annual variability and long-term trends. In this study, bioluminescence was measured at two fixed stations on a daily long term basis: one in San Diego Bay (SDB) for 4 years (1992-1996) and the other for 2.5 years (1993-1996) at San Clemente Island (SCI), located 100 km off the California coast. Additional surface and at-depth bioluminescence data have been collected on a monthly and quarterly basis at both fixed stations and from a research vessel to provide a link between coastal and offshore waters. Additional factors such as seawater temperature, salinity, beam attenuation, and chlorophyll fluorescence were measured. Plankton collections were made weekly in SDB and monthly at SCI. This study provides unique correlated coastal and open ocean data collected on a long-term basis (Figure 1).

2. Methods and materials

2.1 Bioluminescence measurements

Two defined excitation moored bathyphotometers (MOORDEX, University of California, Santa Barbara) were used in San Diego Bay (SDB) and at San Clemente Island (SCI). Under control of on-board computers, these measured stimulated bioluminescence, flow rate, and seawater temperature hourly. Every hour, seawater was pumped for 120 sec at 7-8 L sec^{-1} for a total volume of approximately 840 - 960 L of seawater through a darkened cylindrical 5 l detection chamber approximately 406 mm long and 127 mm in diameter (Case et al. 1993, Neilson et al. 1995). Bioluminescence, excited by the chamber spanning input impeller, was measured by a PMT receiving light from 46 fiber optics tips lining the chamber wall and expressed as photons sec^{-1} ml^{-1} of seawater.

On monthly transits between SDB and SCI an "on-board" sensor system sampled seawater continuously from 3m below the sea surface from a 50m research vessel, the R/V Acoustic Explorer, measuring bioluminescence, seawater temperature, and salinity (Lapota and Losee 1984, Lapota et al. 1988, 1989). A vertically deployed bathyphotometer capable of measuring bioluminescence, temperature, salinity, beam attenuation, and chlorophyll fluorescence to a depth of 100m was used at 4 month intervals (summer, fall, winter, spring) at various stations in the Bight to examine the seasonal changes in the biological and physical structure of the water column (Lapota et at. 1989). Both systems were calibrated with the luminescent bacteria *Vibrio harveyii* in a Quantalum 2000 silicon-photodiode detector. The detector calibration is traceable to a luminol light standard (Matheson et al. 1984).

2.2 Plankton and seawater analysis

Water and plankton samples were collected at 10 stations within the Bight (Figure 1). Monthly transits were made from March 1994 through June 1996 from SCI to SDB to measure surface (3m depth) bioluminescence and collect plankton and seawater samples to determine Chl *a* content. At SDB, weekly plankton and water samples were taken for 4 years while monthly plankton and water samples were collected at SCI for 2.5 years. Because plankton abundance within SDB is usually high, 10 L water samples were concentrated while 40 l samples were filtered for plankton at SCI. Fifteen-liter water samples were collected and filtered from select bathyphotometer depths on the quarterly stations (10, 20, 30, 40, 50, 70, and 90 m). This was accomplished by discharging the bathyphotometer's effluent from its submersible pump through a 130-m long, 2.54 cm (I.D.) hose into a 15 liter Imhoff settling cone. The bottom of the cone was modified with a valve that allowed water to be filtered into collection cups fitted with 20-μm porosity netting. One liter of seawater (unfiltered) was also collected at the each of these depths and frozen in precleaned polycarbonate bottles for later chlorophyll and nutrient analysis. Plankton samples were preserved in a 5% formalin seawater solution. Bioluminescent dinoflagellates were identified to the species level when possible. Chlorophyll *a* was extracted from the seawater samples using standard methods (APHA 1981) and measured by fluorescence as an estimate of biomass on a Turner Model 112 fluorometer (Sequoia-Turner Corp., Mountain View, CA, U.S.A.) and reported as $\mu g\ L^{-1}$.

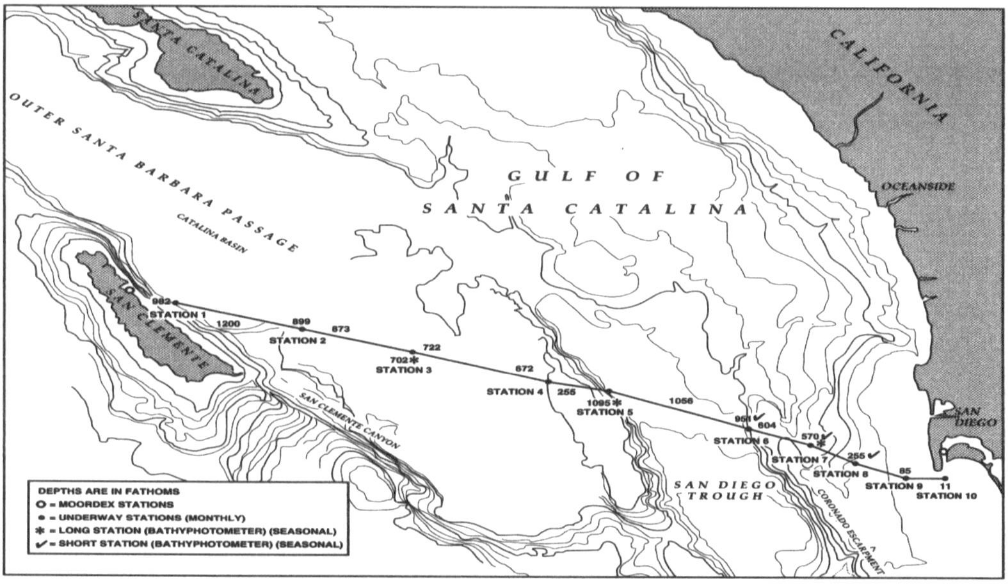

Fig. 1. Bioluminescent study area and cruise track of stations within the Southern California Bight.

2.3 Upwelling, rainfall, and seawater nutrient data bases

Upwelling indices (North Pacific Ocean wind-driven transports) were collected from 1992 through 1996. The indices were computed for 33°N latitude (Schwing et al. 1996) and represent monthly average surface pressure data in cubic meters per second along each 100 m of coastline (Bakun 1973, Eppley 1986). Monthly rainfall data were acquired from the National Weather Service in San Diego. Nutrient and Chl *a* data were accessed from archived CALCOFI data (1992-1996) in the Bight and were averaged along CALCOFI lines 90 and 93 which run west from San Diego to the north and south of San Clemente Island (Hayward et al. 1996). Nitrates (μm L^{-1}) and Chl *a* (μg L^{-1}) along each of the CALCOFI transit lines (stations 93-26 to 93.45 and 90-28 to 90.53) were averaged from the surface to a depth of 50m for 12 cruises conducted from September 1992 through April 1995. These data were used to calculate correlations with bioluminescence, rainfall, and upwelling at SDB.

3. Results

3.1 Mean monthly bioluminescence

Hourly bioluminescence data were averaged for each month. Because minimal bioluminescence was measured during daylight hours, mean monthly values were based on data collected from 2100 h (9:00 P.M.) to 0300 h (3 A.M.) the following day.

Seasonal changes in bioluminescence were observed in SDB. Maximum bioluminescence (1 x 10^8 photons s^{-1} ml^{-1} or greater as a threshold) was measured from March through September for 1993, May through June for 1994, December through May for 1995, and March through April 1996. Minimum bioluminescence (less than 1 x 10^8 photons s^{-1} ml^{-1})

was measured in January through February for 1993, December through February for 1993-94, November for 1994-95, and January through February for 1996 (Figure 2).

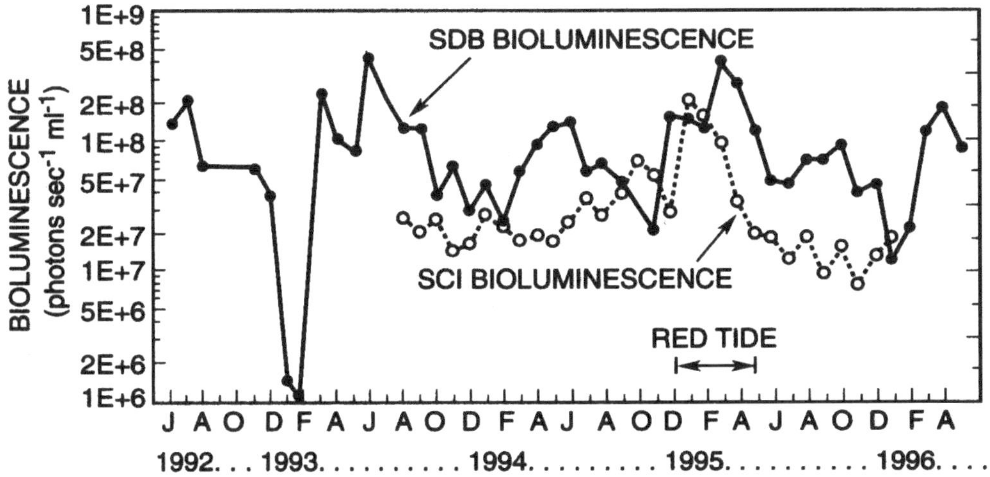

Fig. 2. Mean monthly bioluminescence trends at San Diego Bay and San Clemente Island from 1992-1996.

A red tide of the bioluminescent photosynthetic dinoflagellate, *Gonyaulax polyedra,* developed in the winter of 1994 along the southern California coast and was correlated with an increase in bioluminescence in SDB in December and later at SCI in January through April 1995 (Figure 2). *Noctiluca miliaris* appeared in the plankton collections at SCI following this bloom and produced 57% of the bioluminescence in May 1995. Bioluminescence decreased in SDB during June-July 1995. Mean monthly minimum and maximum bioluminescence in SDB ranged from 1 x 10^6 (February 1993) - 4.5 x 10^8 (June 1993) photons sec^{-1} ml^{-1}. Additionally, during the red tide, bioluminescence averaged 1-2 x 10^8 photons s^{-1} ml^{-1} from December through April 1995, a factor of 10 above the normal measured winter bioluminescence.

In contrast, mean monthly bioluminescence at SCI varied little from August 1993-February 1996 except during the red tide in January 1995 (2 x 10^8 photons s^{-1} ml^{-1}) and persisted through April (Figure 2). Mean monthly bioluminescence ranged from 8 x 10^6 - 3 x 10^7 photons s^{-1} ml^{-1} at SCI.

3.2 Bioluminescent plankton - San Diego Bay

Most bioluminescence in SDB and SCI was emitted by the photosynthetic dinoflagellates *G. polyedra, Ceratium fusus, Pyrocystis noctiluca* as well as from the heterotrophic dinoflagellate, *Noctiluca miliaris*, and several species of *Protoperidinium*. Within SDB, maximum numbers of bioluminescent dinoflagellates (2430 - 17,216 cells L^{-1}) were collected during the spring-summer months while minimal numbers (3 - 2,110 cells L^{-1}) were usually collected in the winter months for 1992, 1993, and 1995 (Figure 3a).

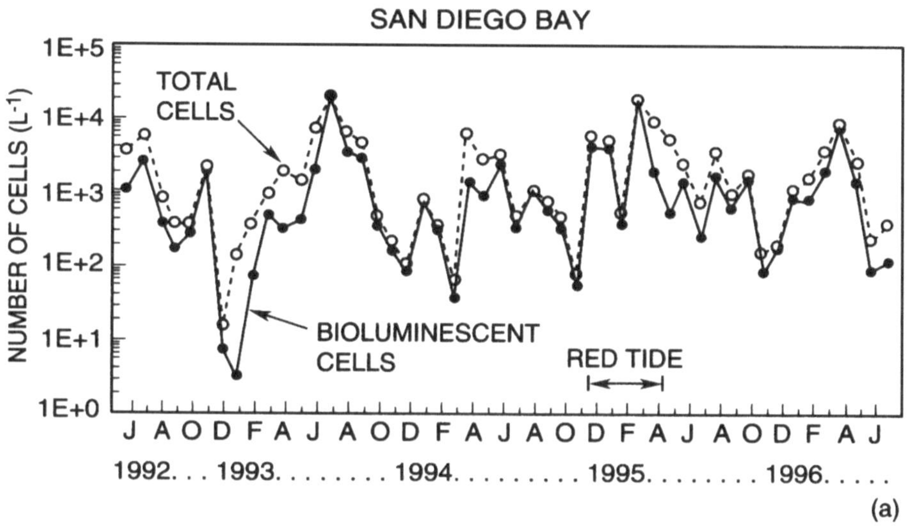

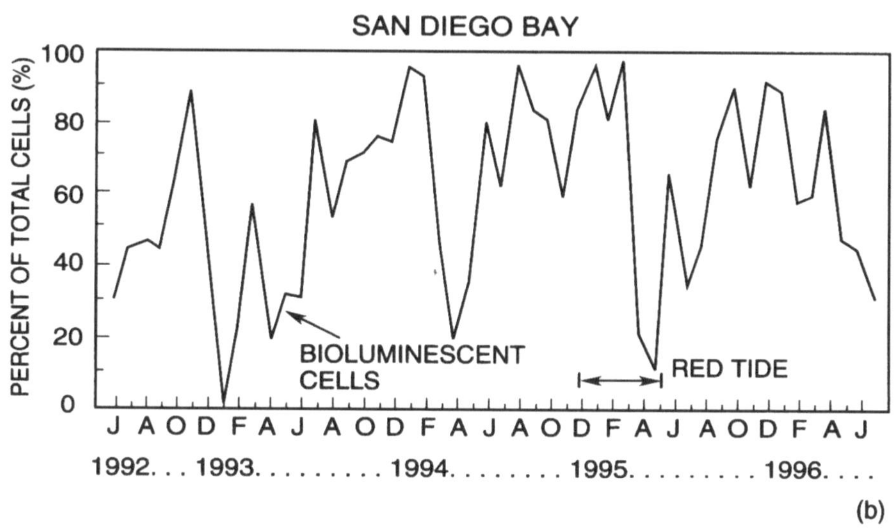

Fig. 3. (a) Total and bioluminescent dinoflagellates collected monthly in San Diego Bay from 1992-1996. (b) Percent of total dinoflagellate cells which are bioluminescent collected in San Diego Bay from 1992-1996.

In most months bioluminescent dinoflagellates represented a substantial percentage of total dinoflagellates (luminous and non-luminous species) (Figure 3b). Of these, *G. polyedra* and *Protoperidinium* spp. were most abundant; found in the winter, spring, and early summer months in SDB (Figure 4). *Gonyaulax polyedra* contributed more than 80% of all luminescent cells from early summer 1993 (Figure 3b). *Ceratium fusus* contributed minimally to the total number of dinoflagellates for the summers of 1993-1994 (Figure 4a). A shift was observed in the dinoflagellate species composition within SDB with *G. polyedra* becoming the dominant species. There were fewer *Protoperidinium* spp. and *C. fusus* at SDB in 1994 (Figure 4).

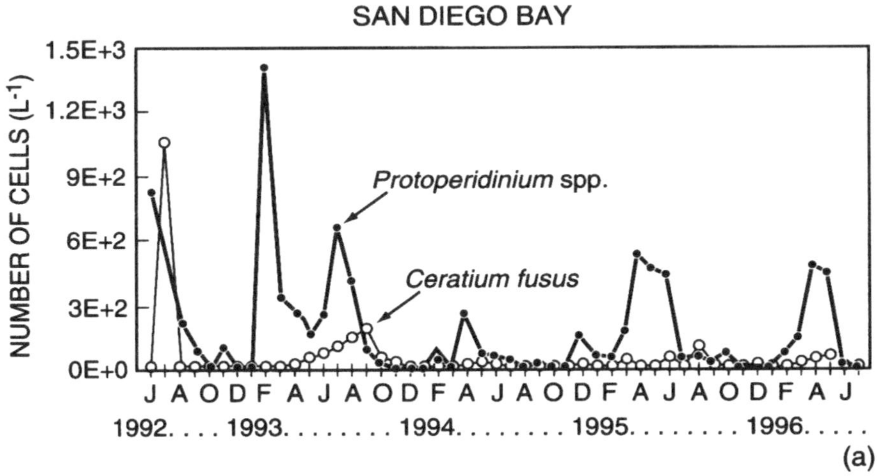

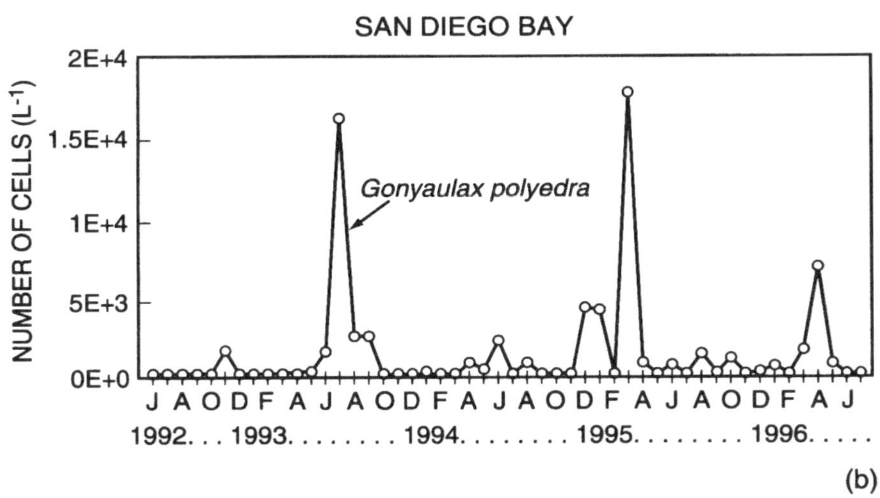

Fig. 4. (a) The numerical abundance of *Ceratium fusus* and *Protoperidinium* spp. each month collected in San Diego Bay from 1992-1996. (b) The numerical abundance of *Gonyaulax polyedra* each month collected in San Diego Bay from 1992-1996.

A light budget was calculated to estimate the contribution of light produced by the various species of bioluminescent dinoflagellates in SDB. Light output from each was measured with a laboratory photometer system by stirring individual cells for 30 sec (Lapota et al. 1989, Lapota et al. 1992). Mean light output for each species was calculated and then multiplied by the number of cells found in each of the monthly plankton samples. The mean light output values for single cells were: *G. polyedra* 1 x 10^8 photons; *C. fusus* 2 x 10^8 photons, *Protoperidinium* spp. 3 x 10^9 photons, *P. noctiluca* 1 x 10^{10} photons, and *N. miliaris* 2 x 10^{10} photons. Photons L^{-1} for each group were plotted from monthly samples collected in SDB (Figure 5). Bioluminescence from each of the groups (% of total bioluminescence) was then estimated (Figure 5). *Protoperidinium* dinoflagellates contributed more than 80% of the bioluminescence in 41% of all months (n = 51 months) and more than 50% of the bioluminescence in 73% of all months. In contrast, *G. polyedra* contributed more than 80% of bioluminescence in SDB in only 2% of all months and more than 50% of all bioluminescence in 18% of all months. *Gonyaulax polyedra* bioluminescence was most pronounced in the fall and winter months. Peaks in *C. fusus* bioluminescence were most pronounced in late summer and fall, however, the contribution to the light budget was minimal. In 1995 and 1996, *G. polyedra* dominance in the winter months was followed by an increase in bioluminescence from *N. miliaris* which attained concentrations of 95 cells L^{-1} in March 1995, 186 cells L^{-1} in April 1995, and 23 cells L^{-1} in May 1995. This same trend and similar cell numbers were encountered in the spring months of 1996.

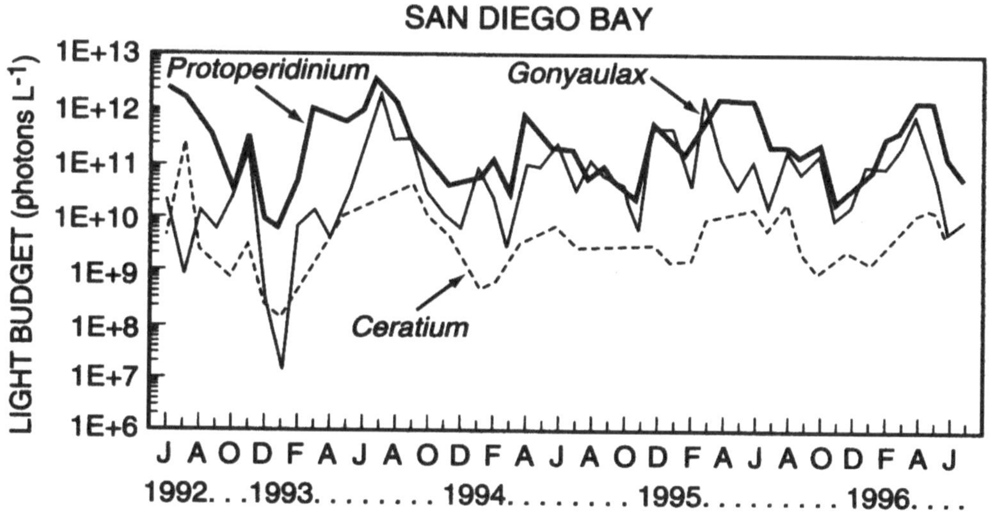

Fig. 5. Bioluminescence produced by each species (photons L^{-1}) monthly in San Diego Bay from 1992-1996.

3.3 Bioluminescent plankton - San Clemente Island

Numbers of luminescent dinoflagellates were lower at SCI than at SDB, ranging from 3 - 211 cells L^{-1} of seawater from August 1993 through December 1994 (Figure 6). The principal species were *G. polyedra* and several species of *Protoperidinium*. The red tide was first observed in January 1995 and persisted through April 1995. Bioluminescence during this event increased approximately 10 times above former levels for both SDB and SCI, although this difference was measured at SDB one month earlier than SCI (Figure 2). Total dinoflagellates and bioluminescent dinoflagellates increased to 16,727 cells L^{-1} and 15,939 cells L^{-1}, respectively at SCI in January 1995 (Figure 7a). Cell numbers remained high through April 1995. *Gonyaulax polyedra* was the predominant red tide bioluminescent dinoflagellate, however several species of *Protoperidinium* increased to numbers as high as 674 cells L^{-1} in February 1995 (Figure 7a). At SCI, bioluminescent dinoflagellates represented a major percentage of all dinoflagellates collected (Figure 7b). The light budget analysis indicated that the species of *Protoperidinium*, again, produced most of the bioluminescence, followed by *Gonyaulax* and *Ceratium* species (Figures 8a,8b). At SCI, *Protoperidinium* contributed more than 80% of all bioluminescence for 60% of all months (n = 30 months) and more than 50% of all bioluminescence for 77% of all months. In contrast, *Gonyaulax* contributed 80% of all bioluminescence for just 1 month (3.3% of all months) and 50% of all bioluminescence for only 10% of the months. During the red tide encountered in the winter and spring of 1995, *G. polyedra* contributed 59%, 42%, 58%, 48%, and 27% of all bioluminescence for the months of January through May 1995, respectively (Figure 8b). As in SDB, *N. miliaris* appeared (~2 cells L^{-1}) following the bloom of *G. polyedra* and produced 57% of the bioluminescence in May 1995 (Figure 8b). The open ocean bioluminescent dinoflagellate, *Pyrocystis noctiluca,* was also found in monthly collections. *Protoperidinium* spp.were present in greater numbers in the spring and summer months; while *G. polyedra* became more prevalent in the fall and winter months.

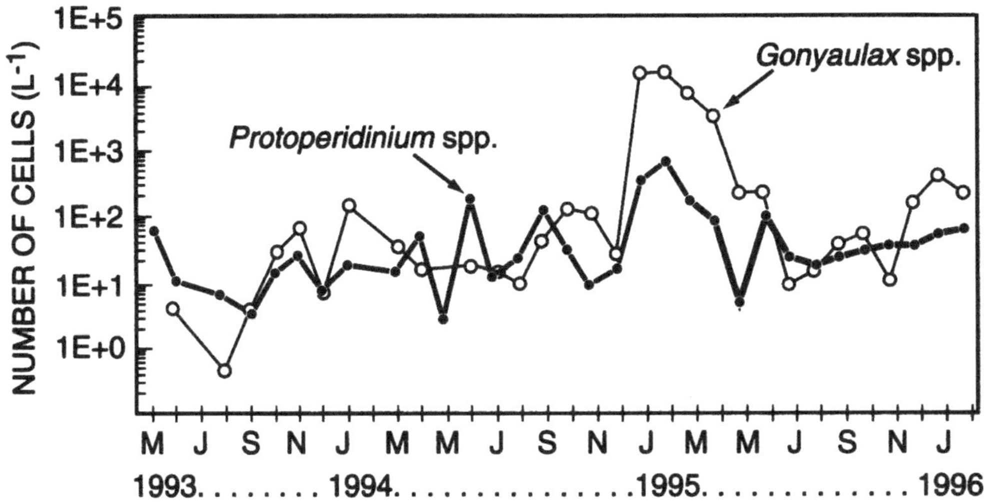

Fig. 6. The abundance of *Protoperidinium* spp. and *Gonyaulax* spp. monthly at San Clemente Island from 1993-1996.

The light budget analysis (photons L^{-1}) at SDB and SCI correlated with measured bioluminescence on a daily (photons ml^{-1} day^{-1}) and monthly (photons ml^{-1} $month^{-1}$) basis, although the light budgets were highly correlated with the later. The light budget analysis reinforces our understanding of which bioluminescent species contributed to measured bioluminescence. SCI had the higher correlations ($r = 0.876$; $p < 0.001$). This may reflect a more constant plankton assemblage over time in contrast to a more variable bay environment where tidal flow into and out of the bay may cause more variation in the populations sampled (Figure 8)

The change in dinoflagellate species composition between summer and fall is well shown in the monthly transits where surface (3-m depth) plankton samples were collected at 10 stations each month from July through October 1994 (Figure 9). We observed a gradual shift in the ratio of *Protoperidinium* spp. to *G. polyedra* between September and October (Figure 9). This trend was also observed in bathyphotometer profiles (0-90 m) for the July and November 1994 stations across the Bight (Figure 10), with *G. polyedra* in both instances dominating in the winter months.

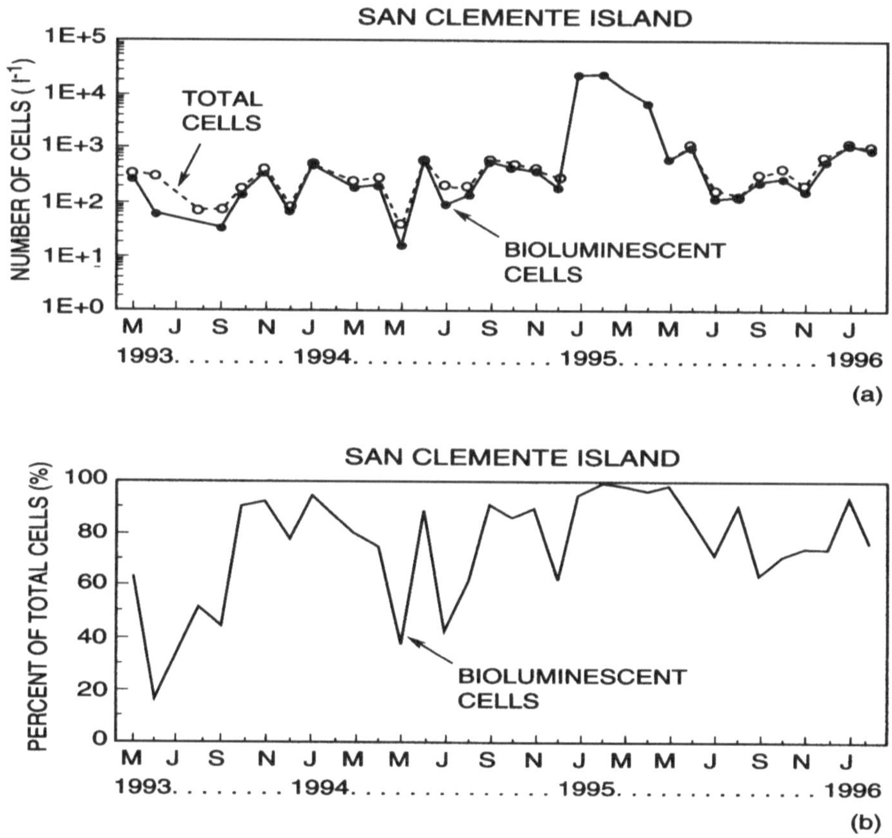

Fig. 7. (a) Total and bioluminescent dinoflagellate cells collected monthly at San Clemente Island from 1993-1996. (b) Percent of total dinoflagellate cells that are bioluminescent monthly at San Clemente Island from 1993-1996.

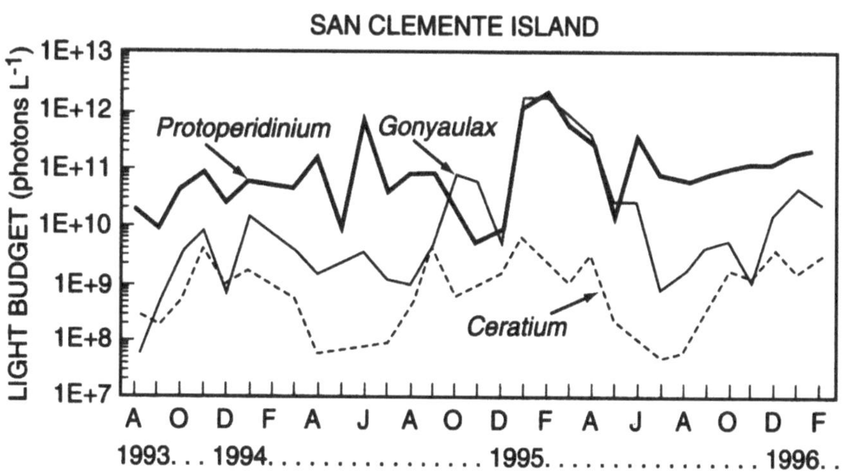

Fig. 8. Bioluminescence produced by each species (photons L^{-1}) monthly at San Clemente Island from 1993-1996.

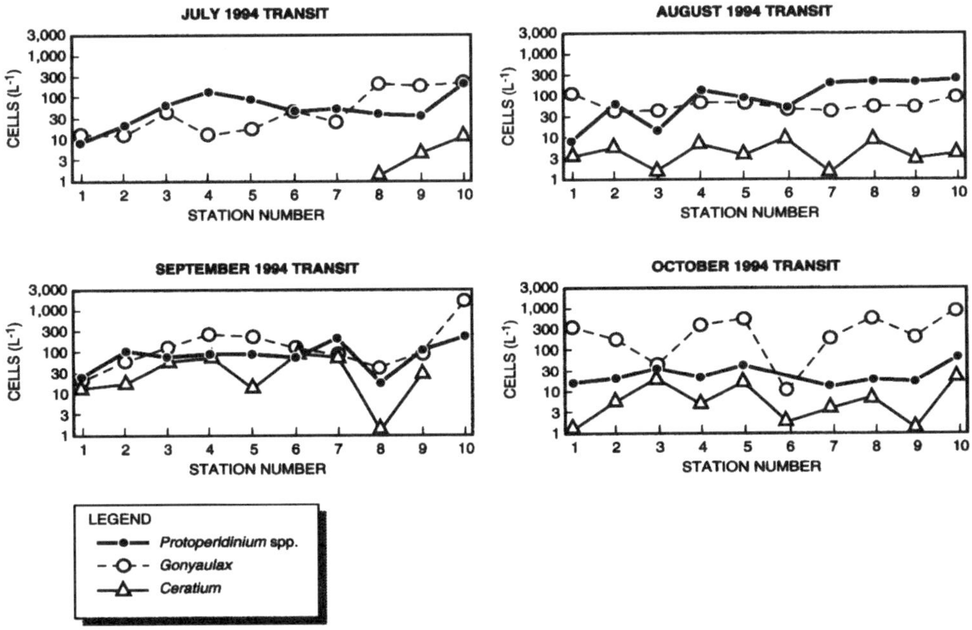

Fig. 9. Dinoflagellate species trends for San Clemente Island to San Diego Bay transits from July 1994 to October 1994.

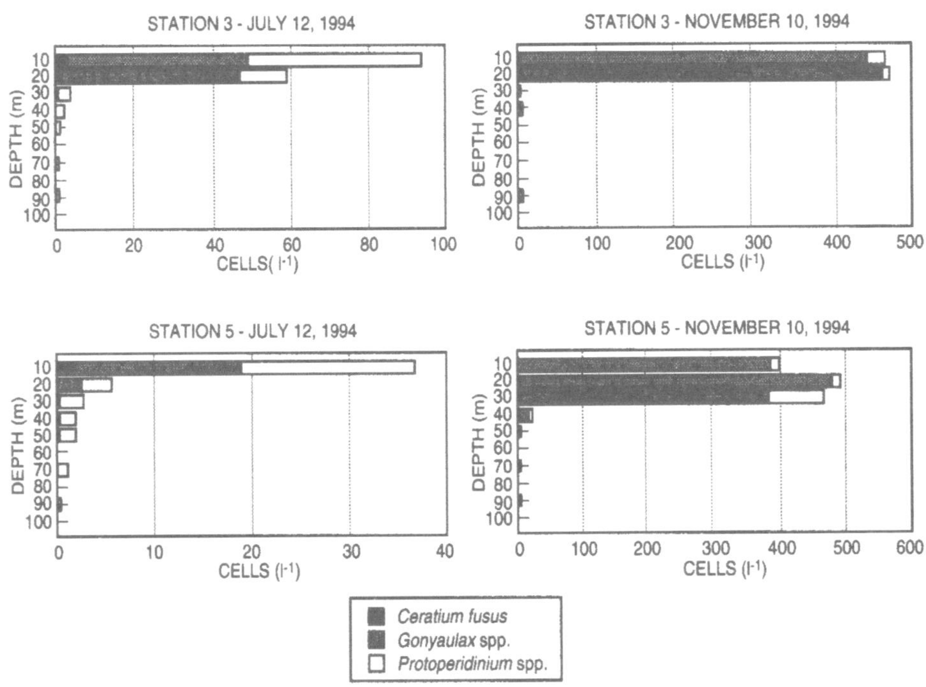

Fig. 10. Dinoflagellate species trends for San Clemente Island to San Diego Bay transits from July 1994 to October 1994.

3.4 Bioluminescence, temperature, rainfall, and chlorophyll relationships

Seawater temperatures in SDB generally ranged from 14°C to 23°C (Figure 11) while mean monthly temperatures at SCI ranged from 14.7°C to 20°C (Figure 12). Correlation coefficients were calculated for both sites. No significant correlations were detected (SDB: $n = 44$ months; $r = 0.131$; $p > 0.10$) (SCI: $n = 30$ months; $r = -0.241$; $p > 0.10$).

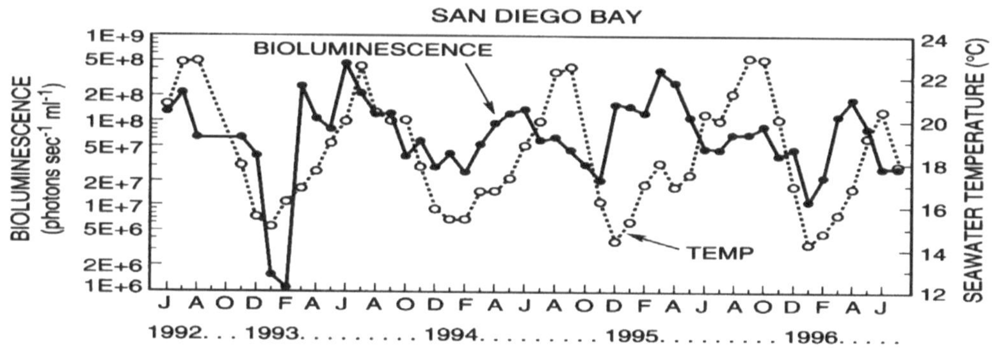

Fig. 11. Mean monthly bioluminescence (photons sec^{-1} ml^{-1}) and seawater temperature (°C) at San Diego Bay from 1992-1996.

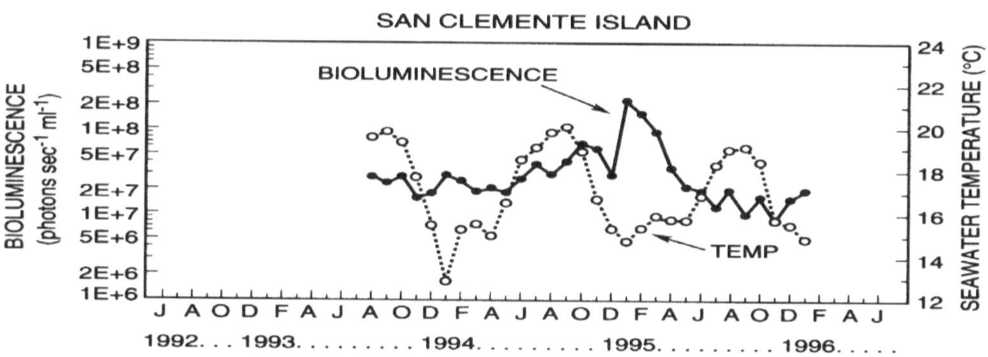

Fig. 12. Mean monthly bioluminescence (photons sec^{-1} ml^{-1}) and seawater temperature (°C) at San Clemente Island from 1993-1996.

3.5 Upwelling indices

Upwelling Indices for 33°N latitude were compared to mean monthly bioluminescence for SDB (Schwing et al. 1996). These indices are north Pacific wind-driven transports computed from monthly average surface pressure data in cubic meters of water per second along each 100 meters of coastline (Figure 13). Cold, nutrient rich water containing nitrates and trace metals are brought to the surface as waters are pushed away from the coast. Nitrates are limited in surface waters (Holm-Hansen et al. 1966, Armstrong et al. 1967, Strickland 1968), and are utilized by all phytoplankton for growth (Spencer 1954, Dugdale 1967, MacIsaac and Dugdale 1969). Inspection of the data shows there is a general trend for upwelling and bioluminescence to co-occur during the same months for 1993-1994 (February - November) but not for 1995 and 1996. Correlation coefficients were not significant for both years (1993:

n = 8 months; r = 0.307; p > 0.10), (1994: n = 8 months; r = 0.617; p > 0.10) and all 4 years (1992-1996: r = 0.111; p > 0.10; n = 47 paired monthly points). Bioluminescence actually began to increase in the winter of 1994 before the onset of upwelling (Figure 13) which suggests that some other factor than upwelling may be controlling the onset of maximum bioluminescence.

3.6 Rainfall effects

Total rainfall for San Diego County from 1992 through 1996 correlated with SDB bioluminescence. Four years of data (1992-1996) showed that years with the greatest precipitation also had measurably more bioluminescence at SDB (Figure 14). The year 1995 was different than prior years in that rainfall preceded upwelling (Figure 13) with a marked increase in bioluminescence. In addition, years with less monthly rainfall (3-4 inches per month for 1994 and 1996 vs. 8-9 inches per month for 1993 and 1995) exhibited less bioluminescence. Total bioluminescence (photons ml^{-1} $year^{-1}$) at San Diego Bay was positively correlated with total rainfall (inches $year^{-1}$) for all 4 years (r = 0.908; n = 4;

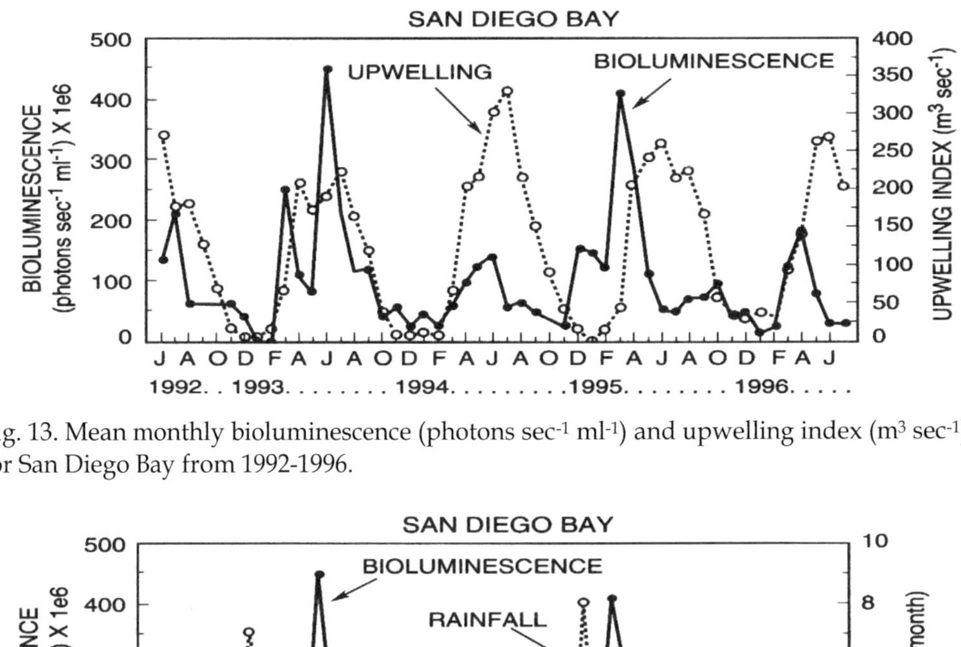

Fig. 13. Mean monthly bioluminescence (photons sec^{-1} ml^{-1}) and upwelling index (m^3 sec^{-1}) for San Diego Bay from 1992-1996.

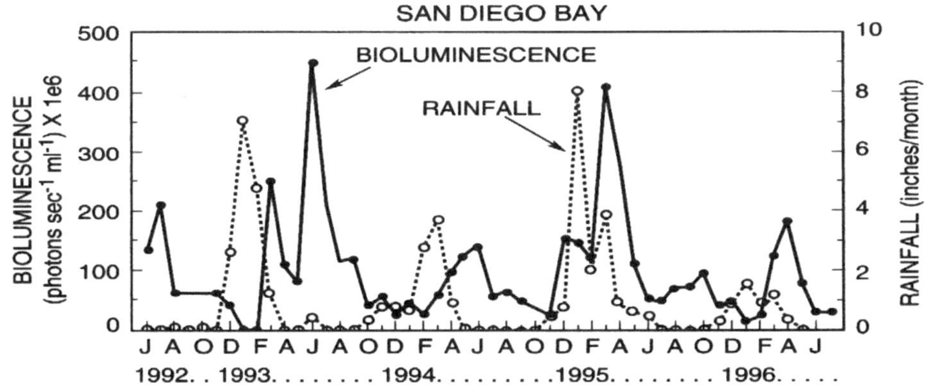

Fig. 14. Mean monthly bioluminescence (photons sec^{-1} ml^{-1}) and monthly rainfall (inches $month^{-1}$) for San Diego Bay from 1992-1996.

$p < 0.05$) (Figure 15). Total bioluminescence at SDB was approximately 41% more in 1992-1993 than in 1993-1994 while 1994-1995 was approximately 66% and 82% greater than in 1993-1994 and 1995-1996, respectively (Figure 16). Bioluminescence at SCI was also 287% greater in 1994-1995 than in 1993-1994. These data suggest that the development of bioluminescence were favored by the consequences of rainfall such as storm runoff nutrients from soil. During the period of 1992 through 1996, of the 77 rain events within San Diego County, 51 events or 66% of all rain events with > 0.1 inch of rainfall were associated with a 50% increase in bioluminescence within three days of the start of rainfall at SDB. Past monitoring programs in San Diego County have shown that storm runoff entering coastal and bay waters during the winter and spring months contains high levels of nitrates and phosphates as well as other nutrients and metals (City of San Diego Stormwater Monitoring Program 1994-1995). CALCOFI data sets also show elevated nitrate levels in surface waters off San Diego along transit lines 90 and 93 following peak rainfall periods.

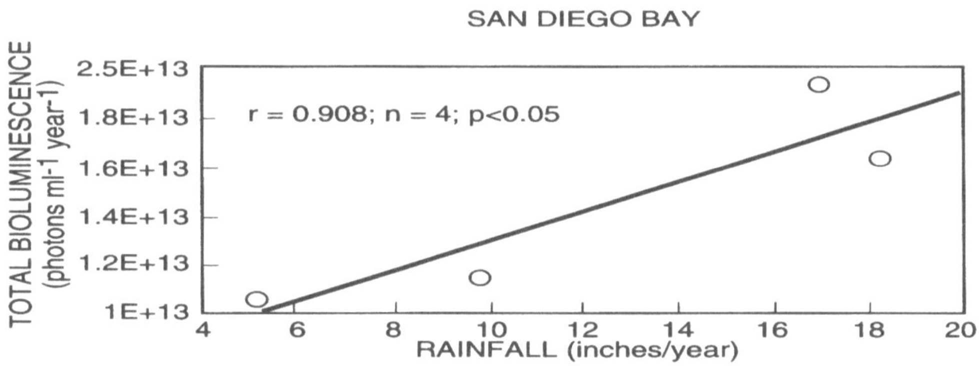

Fig. 15. Correlation of total bioluminescence (photons ml^{-1} year^{-1}) and total rainfall (inches year^{-1}) for San Diego Bay from 1992-1996. (r = correlation coefficient; p = significance level).

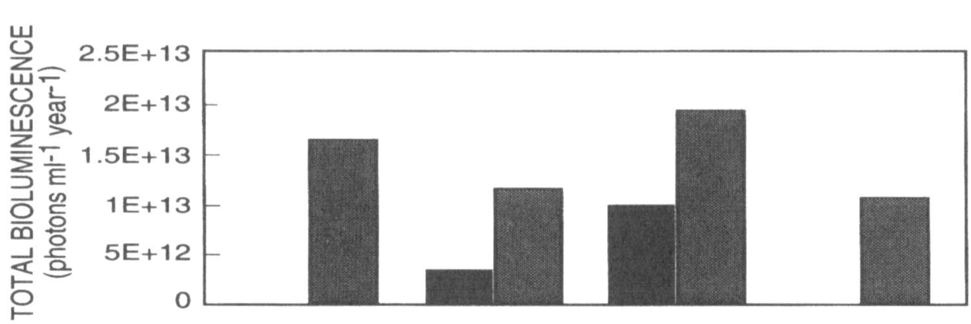

	1992-1993	1993-1994	1994-1995	1995-1996
SCI		3.45E+12	9.91E+12	
SDB	1.64E+13	1.16E+13	1.93E+13	1.06E+13

Fig. 16. Total bioluminescence (photons ml^{-1} year^{-1}) measured at San Diego Bay and San Clemente Island from 1992-1996.

3.7 Seasonality of bioluminescence and chlorophyll *a* at San Clemente Island and San Diego Bay

Total bioluminescence (photons ml^{-1} $year^{-1}$) for each year at SCI and SDB was divided into seasons (Summer, Fall, Winter, and Spring). Total bioluminescence for each month was summed and that subtotal was divided by the entire year's bioluminescence. Seasonal bioluminescence was calculated for SCI (Figure 17a) and SDB (Figure 17b). In 1993-1994, the percent of annual bioluminescence was fairly evenly distributed among all seasons, although the maximum percent of bioluminescence was measured in the winter months at SCI. A winter maximum was again measured the following winter (Figure 17a). For three of the four years at SDB, the maximum percent of annual bioluminescence was measured in the spring. Spring percentages ranged from approximately 30 - 50% of all bioluminescence measured for each of the years (Figure 17b).

Seasonal mean Chl *a* maxima were measured in the spring and winter for 1993-1994 and 1994-1995 at SCI, respectively (Figure 18a). Seasonal mean Chl *a* maxima were measured in the Spring from 1993-1996 for SDB (Figure 18b). Chl *a* was usually greater for all seasons at SDB than at SCI.

The seasonal percentages of bioluminescence for SCI and SDB for all years were averaged as were the seasonal means of Chl *a* for SCI and SDB for all years. Peak bioluminescence was measured in the winter at SCI and the spring at SDB. Forty-four percent of all bioluminescence measured at SCI was in the winter while only 16.5% of the year's total was measured in the summer (Figure 19a). Thirty seven percent of all bioluminescence measured at SDB was in the spring while in the fall, only 14% of the total bioluminescence was measured. Maximum mean Chl *a* was also measured in the winter (0.87 µg L^{-1}) at SCI as was bioluminescence while maximum mean Chl *a* in SDB was measured in the spring (2.39 µg L^{-1}) (Figure 19b). San Clemente Island can then be characterized as having winter maxima for bioluminescence and Chl *a* while San Diego Bay has spring maxima for both.

High levels of chlorophyll at SDB generally occurred either in the spring or summer, although an extended winter - spring peak was measured in January and April-May 1993. At SCI, measured Chl *a* ranged from a low of 0.04 µg L^{-1} in November 1993 to a high of 1.9 µg Chl L^{-1} measured in January 1995 which was probably attributable to the presence of high numbers of *Gonyaulax* , as well as chain diatoms and pico-plankton. Chlorophyll *a* levels were still high through June 1995 although decreasing through the remaining months. Simple correlations were calculated between monthly means of Chl *a* and bioluminescence at SDB and SCI. The correlation between monthly Chl *a* and bioluminescence at SDB was not significant ($r = 0.277$; $n = 43$; $p < 0.10$). At SCI, the correlation was highly siginificant when red tide data were included ($r = 0.88$; $n = 26$; $p < 0.001$). However, when the red tide data were deleted (January - March 1995), the correlation was similar to that at SDB and was not significant ($r = 0.237$; $n = 23$; $p > 0.10$).

Data from the bathyphotometer stations (n= 26) for all six cruises into the Bight (July and November 1994, February, June, November 1995 and March 1996) did not provide a significant correlation of water column bioluminescence with Chl *a* (0 - 90 m) ($r = 0.392$; $p > 0.10$) (Figure 20). Further, integrated water column data for all stations from 1994-1996 or station averages also did not display an association between bioluminescence and Chl *a* (Figure 20).

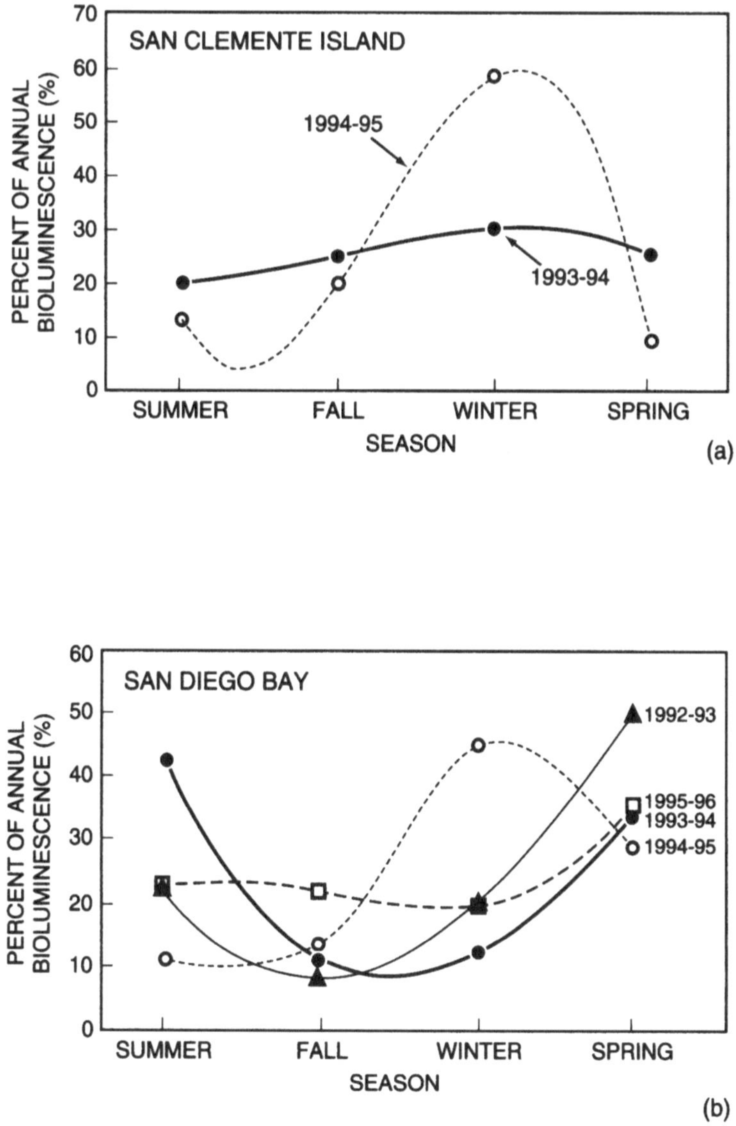

Fig. 17. (a) Distribution of bioluminescence (% of annual bioluminescence) by season measured at San Clemente Island from 1983-1995. (b) Distribution of bioluminescence (% of annual bioluminescence) by season measured in San Diego Bay from 1992-1996.

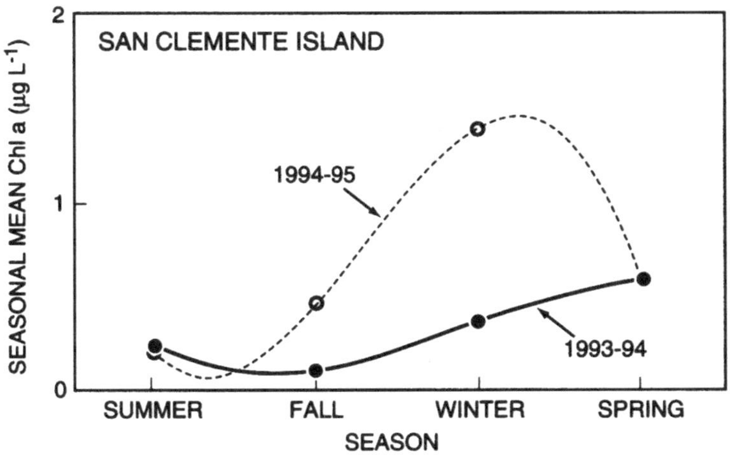

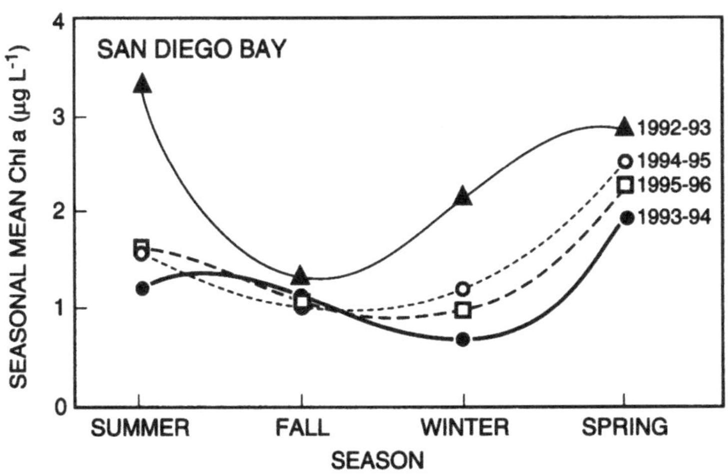

Fig. 18. (a) Seasonal mean Chl *a* (µg L^{-1}) distribution at San Clemente Island from 1993-1995. (b) Seasonal mean Chl *a* (µg L^{-1}) distribution in San Diego Bay from 1992-1996.

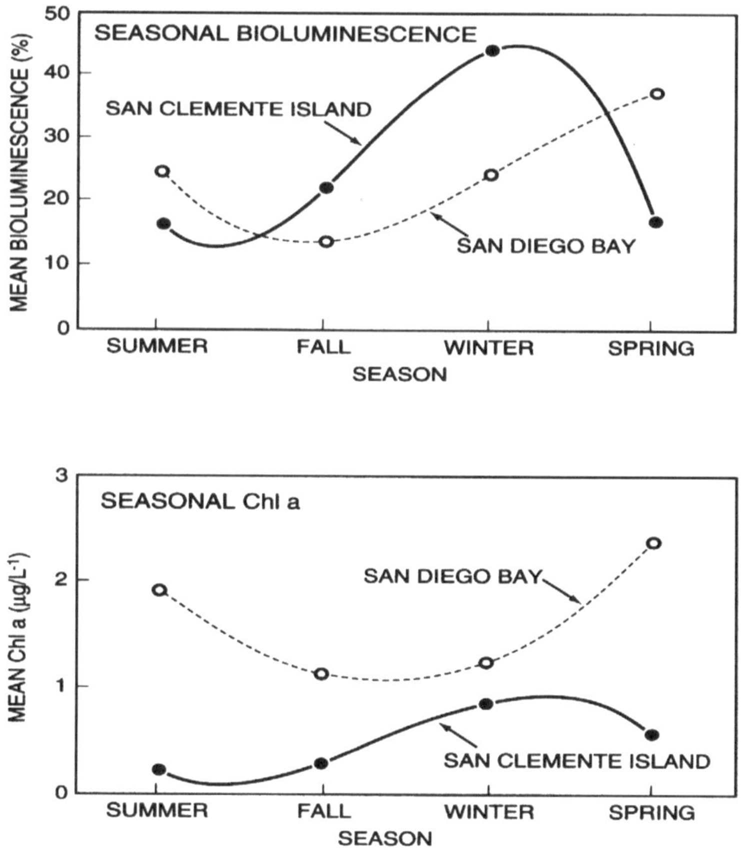

(b)

Fig. 19. (a) Mean seasonal bioluminescence (% of toal) for all years at San Clemente Island and san Diego Bay. Mean seasonal Chl *a* (μg/L) distribution for all years at San Clemente Island and San Diego Bay. (b) Mean seasonal Chl *a* (μg/L) distribution for all years at San Clemente Island and San Diego Bay.

Both the mean integrated water column bioluminescence and Chl *a* increased during the red tide in February 1995 (range: 6×10^{16} - 4×10^{17} photons s^{-1} m^{-2} and 16 - 112 mg Chl m^{-2}, respectively) (Figure 20) in comparison to previous measurements conducted in November 1994 (range: 7×10^{15} - 2×10^{16} photons s^{-1} m^{-2} and 20 - 31 mg Chl m^{-2}, respectively). Mean integrated water column bioluminescence in June 1995 returned to former levels measured in July and November 1994. Station averages of integrated bioluminescence increased as the Southern California mainland was approached. Minimum station averages of integrated bioluminescence were measured at stations 2 and 3, east of SCI in the outer Santa Barbara Passage. Maximum station averages were measured at stations 7, 8 and 9, west of the Coronado Escarpment. Water depths here are the shallowest of all stations measured. Station 7 is located in waters with a depth of 1050 meters while station 9 is located in waters with a depth of 155 meters. Stations 2 and 3 are found in waters with depths of 1660 meters and 1290 meters, respectively.

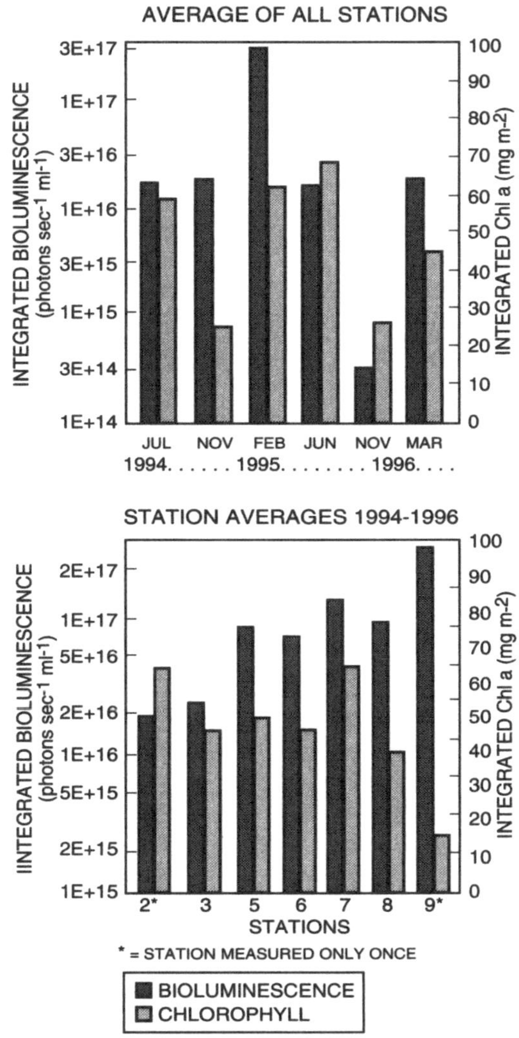

Fig. 20. Integrated bioluminescence (photons sec^{-1} m^{-2}) and Chl *a* (mg m^{-2}) averages of all stations from 1994-1996. Averages of individual stations are also shown.

The vertical structure within the water column with respect to other measured parameters (temperature, percent light transmission, and in vivo Chl fluorescence) changed seasonally (Figure 21). For example, at Station 3 in July (7/12/94), bioluminescence was significantly correlated with *in vivo* Chl fluorescence ($r = 0.673$; $p < 0.001$), beam attenuation ($r = 0.747$; $p < 0.001$), and temperature ($r = 0.892$; $p < 0.001$; Figure 21). Beam attenuation was positively correlated with *in vivo* Chl fluorescence ($r = 0.831$; $p < 0.001$). Maximum bioluminescence and *in vivo* Chl fluorescence were measured at the bottom of the mixed layer (20 m below the sea surface). The mixed layer deepened in November (11/10/94) as did maximum bioluminescence and Chl fluorescence. The correlation between bioluminescence and chlorophyll fluorescence ($r = 0.483$; $p < 0.001$) in November was significant as was bioluminescence and beam attenuation ($r = 0.954$; $p <0.001$) and bioluminescence with temperature ($r = 0.889$; $p < 0.001$). Bioluminescence and *in vivo* Chl fluorescence were not correlated February 12, 1995 at station 3 ($r = 0.062$; $p > 0.10$), however, bioluminescence was still significantly correlated with temperature ($r = 0.765$; $p < 0.001$). On June 13, 1995, bioluminescence and *in vivo* Chl fluorescence were significantly correlated ($r = 0.582$; $p < 0.001$). Bioluminescence and beam attenuation ($r = 0.788$; $p < 0.001$) and bioluminescence and temperature ($r = 0.703$; $p < 0.001$) were significantly correlated. Beam attenuation was significantly correlated with *in vivo* Chl fluorescence ($r = 0.942$; $p < 0.001$). These correlations improved as the mixed layer became shallower in June (Figure 21) in comparison to the deeper mixed layer observed in November 1994 and February 1995 (Figure 21).

4. Discussion

The data show that bioluminescence changes seasonally in the Southern California Bight coastal waters with a maximum and minimum signal in the spring and fall in SDB (Figures 2,14,17b,19a). A winter maximum and summer minimum in bioluminescence was measured at SCI (Figure 2, 17a). In SDB and SCI, we observed a change in the dinoflagellate species composition over a year and its contribution to bioluminescence. We also observed a seasonal change in species composition (summer to winter) at SCI and within the bight (Figures 8a, 9, 10). Chlorophyll *a* also showed similar seasonal trends with respect to location (Figures 18,19). However, measured monthly means of bioluminescence did not correlate with Chl *a* either at SDB or SCI. Mean monthly surface seawater temperature did not correlate with mean monthly bioluminescence at either site; that is, maximum bioluminescence did not always correlate with either maximum or minimum seawater temperatures (Figures 11, 12), although minimum bioluminescence was measured during the coolest water temperatures (winter) at SDB in 1994 and 1996. The largest peak in bioluminescence measured at SCI (winter 1995) was associated with the coolest seawater temperatures (14-15°C) during winter (Figure 12). Coolest water temperatures did not correlate with the upwelling index as maximum indices for 33°N latitude, 119°W longitude generally occurred in June of each year.

Total bioluminescence (photons ml^{-1} $year^{-1}$) was always greater at SDB than at SCI. Total bioluminescence at SDB ranged from 1.06×10^{13} to 1.93×10^{13} photons ml^{-1} $year^{-1}$ (measured from 2100 to 0300 hrs each day) while total bioluminescence measured at SCI was from 3.45×10^{12} to 9.91×10^{12} photons ml^{-1} $year^{-1}$. In 1993-1994, 3 times more bioluminescence was measured at SDB than at SCI. These differences lessened to a factor of 2 in 1994-1995 between both sites when a massive bioluminescence red tide was observed to extend south from Santa Barbara, California to Ensenada, Mexico and 100km offshore to SCI. At times, monthly differences in total bioluminescence were 8 times greater at SDB than at SCI in Spring 1994 and 1995.

Bioluminescent dinoflagellates, in most instances, comprised most of the dinoflagellates collected at SDB and SCI (Figures 3, 7). In SDB, bioluminescent dinoflagellates made up at least 80% of all dinoflagellates. Numbers of bioluminescent dinoflagellates dropped noticeably in the winter and spring at SDB (< 30% of total dinoflagellates) at SDB. Decreases in bioluminescent dinoflagellates were observed at SCI in late spring at SCI. The bioluminescent dinoflagellate assemblage at both SDB and SCI was composed of *Ceratium*, *Gonyaulax*, *Protoperidinium*, and *Noctiluca* species. *Pyrocystis noctiluca* was a recurring species

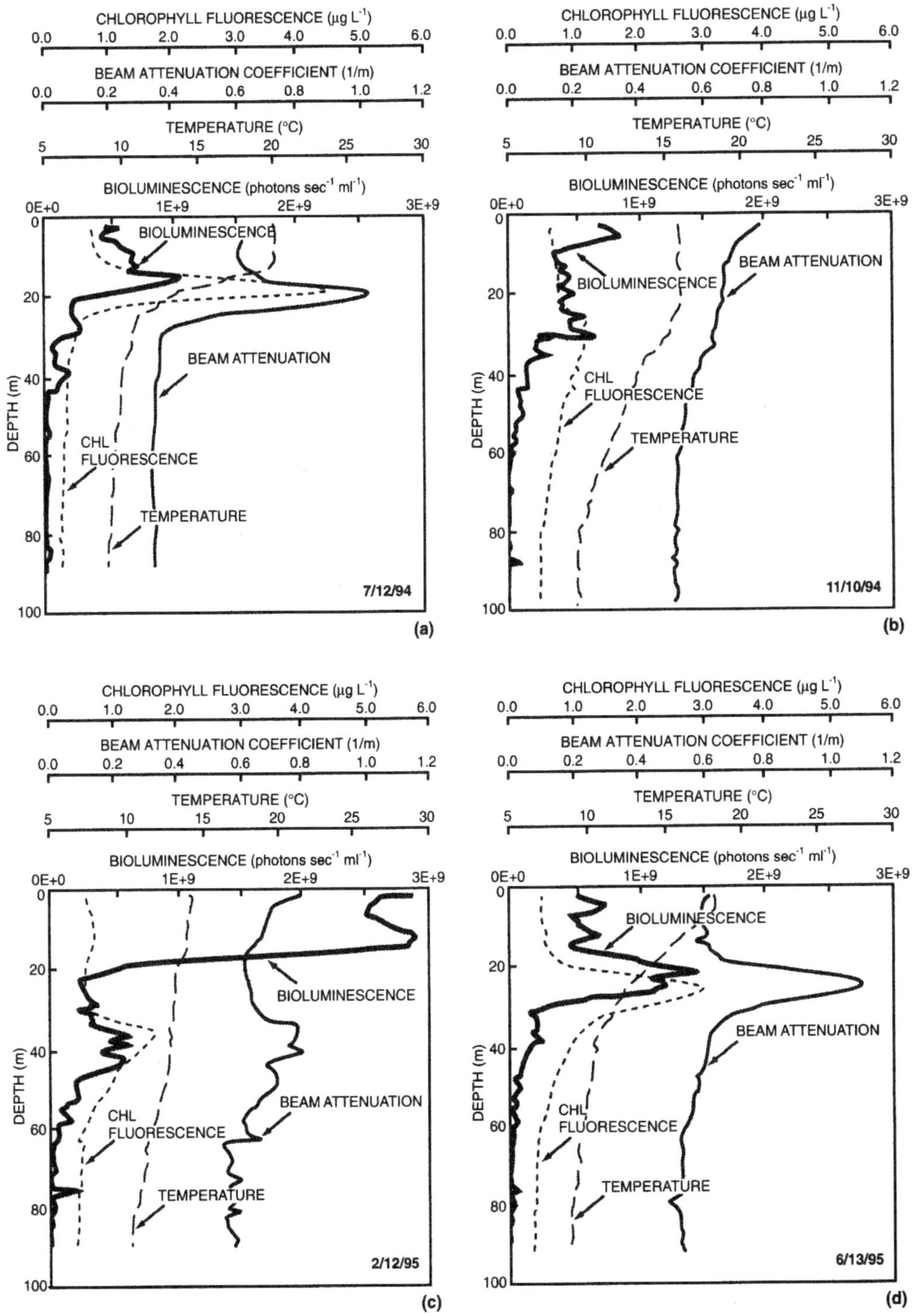

Fig. 21. Bathyphotometer profiles at station 3 for (a) July 12, 1994; (b) station 3 November 10, 1994; (c) station 3 February 12, 1995; (d) station 3 June 13, 1995.

found at SCI. *Protoperidinium* spp. and *Gonyaulax polyedra* contributed most of the bioluminescence at both sites. *Noctiluca miliaris* contributed substantial bioluminescence following increases in *G. polyedra* at SDB in 1995 and 1996 and at SCI in 1995.

Total rainfall was significantly correlated with measured bioluminescence at SDB ($r = 0.908$; $n = 4$; $p < 0.05$). Years with the greatest rainfall (1993, 1995) affected the total bioluminescence which implies that processes associated with rainfall, such as storm water runoff may be stimulating dinoflagellate and algal production in coastal waters (Anderson 1964; Eppley et al., 1978). We observed that the upwelling index did not directly correlate with SDB bioluminescence unless the index was shifted back 1 month ($r = 0.476$; $p < 0.001$). However, if mean monthly bioluminescence was shifted forward 2 months with mean rainfall, a significant correlation was observed ($r = 0.472$; $p < 0.01$). The upwelling index and nitrates ($\mu m\ L^{-1}$) measured in coastal waters, were significantly correlated when nitrate levels were shifted forward in time 1 month ($r = 0.679$; $p < 0.001$). We must then assume that some other factor besides upwelling is providing a stimulatory effect to dinoflagellate bioluminescence. Multiple regression analysis showed that rainfall, upwelling, and temperature were the most important conditions to predict bioluminescence and that when rainfall was moved ahead in time by 2 months, we could account for 24.7% of the observed variance to predict bioluminescence from 1992 - 1996 ($R^2 = 0.2468$; $F=2.469$; $p<0.05$). Increased nitrate levels were observed in coastal waters beyond SCI during the winter months and spring months; before maximum upwelling. The source of these nitrates may be in storm water runoff. Support for "new sources of nitrogen" versus "recycled nitrogen" and other nutrients entering the water column is not new. Some studies have shown that river inputs into the ocean can carry high levels of nutrients needed for algal growth (Harrison 1980, Fogg 1982, Mooers et al. 1978, Lalli and Parsons, 1993). Others have found that ferric iron is a limiting nutrient for phytoplankton growth (Menzel and Ryther 1961) and that high levels of iron are often associated with river runoff (Williams and Chan 1966). Iron is needed by phytoplankton to utilize nitrates for growth (Ryther and Kramer 1961). The availability of iron is enhanced by chelation with dissolved organic matter. That is, organically bound iron from storm runoff may stimulate the growth of phytoplankton (Kawaguchi and Lewitus 1996). Similarly, elevated phytoplankton levels off Del Mar, California following storm water runoff were attributed to increased nutrient inputs from land (Eppley et al. 1978).

The bathyphotometer stations showed that bioluminescence and Chl fluorescence were positively correlated, during the summer months, when the water column stabilized with a shallow thermocline. These significant positive correlations broke down with water column mixing during the fall and winter months but were reestablished with the development of the thermocline during the spring and summer months. Several species of *Protoperidinium* were the predominant dinoflagellate in the spring and summer months while *G. polyedra* was important during the fall and winter months; not only in surface waters, but at depth. The increased bioluminescence and chlorophyll levels associated with the red tide at SCI are remarkable for their duration since they persisted from January through April 1995. This strengthens the inference that the physical environment in the bight is fairly stable with respect to seasonality, and that bioluminescence is strongly influenced by seasonal rainfall and runoff.

Southern California Bight bioluminescence is similar to that found in coastal waters of Vestfjord, Norway (Lapota 1990, unpublished), and the Arabian Sea (Lapota & Rosenberger 1990), but higher than that found in the Sargasso Sea (Batchelder & Swift 1989), the North Atlantic (Neilson et al. 1995) and the Beaufort Sea (Lapota et al. 1992). The vertical structure of bioluminescence was correlated with Chl fluorescence for some of the stations in the Bight. However, integrations between bioluminescence and chlorophyll were positively correlated, but weak. Strong positive correlations between bioluminescence and chlorophyll fluorescence were observed during the red tide in February 1995. At depth, seawater temperature correlated strongly with the vertical distribution of bioluminescence, as did transmission. In contrast, weaker correlations were observed between bioluminescence and Chl fluorescence. Other studies have infrequently observed correlations which may be dependent on the season the study was conducted (Lapota et al. 1989 Young et al. 1992, Neilson et al. 1995, Ondercin et al. 1995). An obvious conclusion is that the primary dinoflagellates which are contributing much of the bioluminescence do not contain Chl *a*. These would include the heterotrophic *Protoperidinium* dinoflagellates. These dinoflagellates produce as much as 30 times more light per cell than does *G. polyedra* (Biggley 1969, Lapota et al. 1992). This could explain the poor correlations between bioluminescence and chlorophyll. Consequently, these results impact models predicting bioluminescence from global ocean primary production and ocean color (Young et al. 1992, Ondercin 1995) since these are based on the assumption that much of the oceanic bioluminescence is derived from photosynthetic bioluminescent dinoflagellates (Ondercin 1995). It is clear that from this and other studies (Lapota et al. 1989, 1992,1993 a, b, Swift et al. 1995, Neilson et al. 1995) that *Protoperidinium* dinoflagellates dominate surface water bioluminescence in the world's oceans for a significant portion of the year.

5. Acknowledgements

We gratefully acknowledge support by the Office of Naval Research, VA through program element 0601153N-03102 and the Naval Space and Warfare Systems Center, Pacific, CA and Dr. James Case at the University of California, Santa Barbara for his guidance throughout this study. We also thank Connie H. Liu and Joel Guerrero (Naval Space and Warfare Center, Pacific for conducting chlorophyll and chemical analyses and participation in the cruises.

6. References

American Public Health Association (1981). Standard methods for the examination of water and wastewater, 15th ed. Washington, D.C., 1134 pp.

Anderson, G.C.(1964). The seasonal and geographic distribution of primary productivity off the Washington and Oregon Coasts. Limnol. Oceanogr. 9: 284-302.

Armstrong, F.A.J., Stearns, C.R., Strickland, J.D.H. (1967) The measurement of upwelling and subsequent biological processes by means of the Technion Autoanalyser® and associated equipment. Deep-Sea Res.14: 381-389.

Bakun, A. (1973). Coastal upwelling indices, west coast of North America, 1946-71, Nat. Oceanic Atmos. Adm. (US), Spec. Sci. Rep.- Fish No. 671, 103 pp.

Batchelder, H.P., Swift, E. (1989). Estimated near-surface mesoplanktonic bioluminescence in the western North Atlantic during July 1986. Limnol. Oceanogr. 34: 113-128.

Biggley, W.H., Swift, E., Buchanan, R.J., Seliger, H.H. (1969). Stimulable and spontaneous bioluminescence in the marine dinoflagellates, Pyrodinium bahamense, Gonyaulax polyedra, and Pyrocystis lunula. J. Gen. Physiol. 54: 96-122.

Bityukov, E.P., Rybasov, V.P., Shaida, V.G. (1967). Annual variations of the bioluminescent field intensity in the neritic zone of the Black Sea. Oceanology 7 (6): 848-856.

Case, J.F., Widder, E.A., Bernstein, S., Ferer, K., Young, D., Latz M., Geiger, M., Lapota, D. 1993). Assessment of marine bioluminescence. Nav. Res. Rev. 45: 31-41.

Clarke, G.L., Kelly, M.G. (1965). Measurements of diurnal changes in bioluminescence from the sea surface to 2,000 meters using a new photometric device. Limnol. Oceanogr. 10 (supplement), R54-66.

Dugdale, R.C. (1967). Nutrient limitation in the sea: dynamics, identification and significance. Limnol. Oceanogr. 12: 685-695.

Eppley, R.W., Sapienza, C., Renger, E.H. (1978). Gradients in phytoplankton stocks and nutrients off southern California in 1974-76, Estuarine Coastal Mar. Sci. 7: 291-301.

Eppley, R.W. (1986). Plankton Dynamics of the Southern California Bight, Lecture Notes on Coastal and Estuarine Studies, XIII. Eppley, R.W.(ed.), Springer-Verlag, Berlin, 373 pp.

Fogg, G.E. (1982). Nitrogen cycling in sea waters. Phil. Trans. Roy. Soc. Lond. Ser. B, 296: 511-520.

Harrison, W.G. (1980). Nutrient regeneration and primary production in the sea. In: Falkowski, P. (ed.) Primary Productivity in the Sea, Brookhaven Symposium Biology 31, Plenum, New York, pp. 433-460.

Hayward, T.L., Cummings, S.L., Cayan, D.R., Chavez, F.P., Lynn, R.J., Mantyla, A.W., Niiler, P.P., Schwing, F.B., Veit, R.R., Venrick, E.L. (1996). The state of the California Current in 1995: Continuing declines in macrozooplankton biomass during a period of nearly normal circulation. CalCOFI Reports 37: 22-37.

Holm-Hansen, O., Strickland, J.D.H., Williams, P.M. (1966). A detailed analysis of biologically important substances in a profile off southern California. Limnol. Oceanogr. 11: 548-561.

Kawaguchi, T., Lewitus, A.J. (1996). The potential effect of urbanization on iron Bioavailability and the implication for phytoplankton production, International Conference on Shellfish Restoration, South Carolina Sea Grant, Hilton Head, South Carolina, USA, 20-23 November 1996.

Kinnetic Laboratories, Inc (1995). City of San Diego and Co-Permittee Stormwater Monitoring Program 1994-1995, Carlsbad, Calif.

Lalli, C.M., Parsons, T.R. (1993). Biological Oceanography, Pergamon Press, Oxford, England, pp. 141-143.

Lapota, D., Losee, J.R. (1984). Observations of bioluminescence in marine plankton from the Sea of Cortez. J. exp. Mar. Biol. Ecol. 77: 209-240.

Lapota, D., Galt, C., Losee, J.R., Huddell, H.D., Orzech, K., Nealson, K.H. (1988). Observations and measurements of planktonic bioluminescence in and around a milky sea. J. exp. Mar. Biol. Ecol. 119: 55-81.

Lapota, D., Geiger, M.L., Stiffey, A.V., Rosenberger, D.E., Young, D.K. (1989). Correlations of planktonic bioluminescence with other oceanographic parameters from a Norwegian fjord. Mar. Ecol. Prog. Ser. 55: 217-227.

Lapota, D., Rosenberger, D.E. (1990). Bioluminescence measurements and light budget analysis in the western Arabian Sea. EOS, Trans. of the Am. Geophy. Union 71: 97.

Lapota, D., Rosenberger, D.E., Lieberman, S.H. (1992a). Planktonic bioluminescence in the pack ice and the marginal ice zone of the Beaufort Sea. Mar. Biol. 112:665-675.

Lapota, D., Young, D.K., Benstein, S.A., Geiger, M.L., Huddell, H.D., Case, J.F. (1992b). Diel bioluminescence in heterotrophic and photosynthetic marine dinoflagellates in an Arctic Fjord. J. mar. biol. Ass. U.K. 72: 733-744.
MacIsaac, J.J., Dugdale, R.C. (1967). The kinetics of nitrate and ammonia uptake by natural populations of marine phytoplankton. Deep-Sea Res. 16: 45-57.
Matheson, I.B.C., Lee, J., Zalewski, E.F. (1984). A calibration technique for photometers. Ocean Optics 7 (489): 380-381.
Menzel, D.W., Ryther, J.H. (1961). Nutrients limiting the production of phytoplankton in the Sargasso Sea, with special reference to iron. Deep-Sea Res. 7: 276-281.
Mooers, C.N.K., Flagg, C.N., Boicourt, W.C. (1978). Prograde and retrograde fronts. In Bowman, M., Esias, W.E. (eds.) Ocean Fronts in Coastal Processes, Springer, Berlin, pp. 43-58.
Neilson, D.J., Latz, M.I., Case, J.F. (1995). Temporal variability in the vertical structure of bioluminescence in the North Atlantic ocean. J. Geophys. Res. 100 (C4):6591-6603.
Ondercin, D.G., Atkinson, C.A., Kiefer, D.A. (1995). The distribution of bioluminescence and chlorophyll during the late summer in the North Atlantic: Maps and a predictive model. J. Geophys. Res. 100 (C4): 6575-6590.
Ryther, J.H., Kramer, D.D. (1961). Relative iron requirement of some coastal and off-shore plankton algae. Ecology 42: 444-446.
Schwing, F.B., O'Farrell, M., Steger, J., Baltz, K. (1996). Coastal upwelling indices, west coast of North America 1946-1995. NOAA-TM-NMFS-SWFSC-231.
Seliger, H.H., Fastie, W.G., Taylor, W.R., McElroy, W.D. (1961). Bioluminescence in Chesapeake Bay. Science 133: 699-700.
Spencer, C.P. (1954). Studies on the culture of a marine diatom. J. mar. biol. Ass. U.K. 33: 265-290.
Strickland, J.D.H. (1968). A comparison of profiles of nutrient and chlorophyll concentrations taken from discrete depths and by continuous recording. Limnol. Oceanogr. 13: 388-391.
Stukalin, M.V. (1934) Bioluminescence of the Okhotsk Sea. Vest dal'nevost Fil. Akad. Nauk. SSSR 9: 137-139.
Swift, E., Sullivan, J.M., Batchelder, H.P., Van Keuren, J., Vaillancourt, R.D. (1995). Bioluminescent organisms and bioluminescent measurements in the North Atlantic Ocean near latitude 59.5°N, longitude 21°W. J. Geophy. Res. 100 (C4):6527-6547.
Tarasov, N.I. (1956). Marine luminescence. USSR Academy of Sciences, Moscow [in Russ.] [Engl transl by Naval Oceanographic Office (No NOOT-21); National Space Technology Laboratories Station, Bay St. Louis, MS].
Tett, P.B. (1971). The relation between dinoflagellates and the bioluminescence of sea water. J. mar. biol. Ass. U.K. 51:183-206.
Williams, P.M., Chan, K.S. (1966). Distribution and speciation of iron in natural waters: Transition from river water to a marine environment, British Columbia, Canada. J. Fish. Res. Bd. Canada 23: 575-593.
Yentsch, C.S., Laird, J.C. (1968). Seasonal sequence of bioluminescence and the occurrence of endogenous rhythms in oceanic waters off Woods Hole, Massachusetts. J. mar. Res. 26:127-133.
Young, D.K., Lapota, D., Hickman, G.D. (1992). Ocean color effects on the operation of active and passive optical aircraft/satellite systems, Applications of ocean color to Naval warfare, In: Hickman, G.D. (ed.) Applications of Ocean Color to Naval Warfare. Naval Oceanographic and Atmospheric Research Laboratory, Stennis Space center, MS 39529-5004, pp 32-43.

2

Seasonal Changes of Bioluminescence in Photosynthetic and Heterotrophic Dinoflagellates at San Clemente Island

David Lapota
Space and Naval Warfare Systems Center, Pacific
USA

1. Introduction

A significant portion of bioluminescence in all oceans is produced by dinoflagellates. Numerous studies have documented the ubiquitous distribution of bioluminescent dinoflagellates in near surface waters (Seliger et al., 1961; Yentsch and Laird 1968; Tett 1971; Tett and Kelly 1973). The number of bioluminescent species and their relative abundance change temporally, with depth, and geographically. Dinoflagellates are most abundant in coastal waters and inland seas and are less abundant in the open ocean (Colebrook and Robinson, 1965; Dodge and Hart-Jones, 1977). Studies have been conducted to determine the species contributing to bioluminescence. In several studies, this involved making plankton collections, isolation and measurement of cells with a laboratory photometer to quantify the light output of several species of bioluminescent dinoflagellates (Lapota and Losee 1984; Batchelder and Swift 1989; Lapota et al., 1992a,b; Swift et al., 1995). These studies were limited to short sampling periods (days-weeks) and to specific locations. There is also evidence that dinoflagellates undergo changes in light output which may be attributable to environmental conditions. For example, cells of *Protoperidinium* spp. produce more bioluminescence when nutritional requirements were optimized in the laboratory (Buskey 1992; Latz 1993). Others have observed that the bioluminescence potential of a dinoflagellate is related to its surface area or cell volume for several species, which might be related to light and nutrient history (Seliger et al. 1969; Seliger and Biggley 1982; Swift et al. 1995; Sullivan and Swift 1995). Bioluminescence may also be a function of light, temperature, and nutrient history (Sweeney, 1981). Other data have suggested that cells of the same species in the same study display differences in bioluminescence. These observations may indicate that cells are exposed to a wide range of environmental conditions affecting light output on a short time scale such as light history, nutrient history, grazing pressure by herbivores and consequent loss of potential bioluminescent capacity (Swift et al., 1981; Sullivan and Swift, 1995).

Despite strong interest in short term process effects on dinoflagellates there have been few investigations on the seasonality of marine bioluminescence (Tett 1971; Bityukov et al. 1967; Lapota et al. 1997). Long term aspects of the development of bioluminescence are unknown for most oceans. The present study was designed to cast light on this question. A station for measuring bioluminescence was established in August 1993 at San Clemente Island (SCI), 100 km offshore of Southern California. Bioluminescence was measured with a moored bathyphotometer (MOORDEX) hourly through February 1996. Other environmental parameters such as nutrients, chlorophyll, and associated plankton species were measured and collected on a monthly and quarterly basis(Lapota et al. 1997). In the present study, plankton samples were collected and tested for bioluminescence on a quarterly basis to: 1) determine which dinoflagellate species were bioluminescent and 2) observe differences in light output on a seasonal basis. The latter is an important consideration because seasonal changes in bioluminescence from dinoflagellates might possibly indicate a response to regional seasonal environmental changes. These factors include the available nutrients and light for the photosynthetic species (*Ceratium, Gonyaulax*[1], *Pyrocystis*) and the availability of diatoms and smaller algal cells consumed by the heterotrophic *Protoperidinium* dinoflagellates. Seasonal changes in light output will affect the bioluminescence light budget of all species. Published light budgets are limited and specific for limited oceanic areas (Swift et al. 1983, 1985a,b; 1995; Batchelder and Swift 1989; Lapota et al. 1988, 1989, 1992a,b; Buskey 1991) , the number of species tested, or modeled to predict bioluminescence output based on the calculated cell surface area (Seliger and Biggley 1982). This study will complement earlier laboratory work and enlarge these observations by identifying distinct seasonal differences in bioluminescence of open ocean dinoflagellates over a two year period.

2. Methods and materials

2.1 Plankton collections

Plankton samples were collected from the Naval Ordinance Test Station (NOTS) pier at SCI on a quarterly basis from the summer of 1994 through spring 1996. The pier is on the leeward side of SCI and is in a water depth of approximately 15 meters. The island shelf begins to deepen to greater than 100 meters within 150 meters of the pier (Figure 1). A MOORDEX bathyphotometer was suspended under the pier by a 1cm diameter stainless steel cable. Depending on the height of the tide, MOORDEX was usually at a depth of 2-3 meters below the sea surface (Lapota et al. 1994a). A similar MOORDEX bathyphotometer was also deployed in San Diego Bay (SDB) from 1992-1996 for comparative coastal measurements of bioluminescence (Lapota et al., 1997). Plankton samples were always collected in the late afternoon. A 20-µm mesh plankton net and attached collection cup was lowered off the pier to a depth of approximately 10 meters and vertically retrieved. Samples were diluted with fresh filtered (0.45 µm) seawater and transported to a field trailer and kept in a temperature-controlled incubator (Coolatron™) at ambient seawater temperature. Individual cells were viewed microscopically and isolated by pipet and placed in 4-ml spectrophotometric disposable cuvettes with 3-ml of (0.45 µm) filtered seawater. All isolations were completed no later than 1 hour prior to sunset to prevent premature stimulation of bioluminescence. Bioluminescence measurements were always conducted midway into scotophase, about 8-9 hours after collection. Two consecutive nights of isolation and testing were conducted each season (i.e., winter: December 21- March 21; spring: March 22-June 21; summer: June 22-September 21; fall: September 22-December 21).

[1] *Gonyaulax polyedra* has recently been renamed *Lingulodinium polyedrum* Stein (Dodge 1989), but for simplicity we will use the former name because of its wider use in earlier literature.

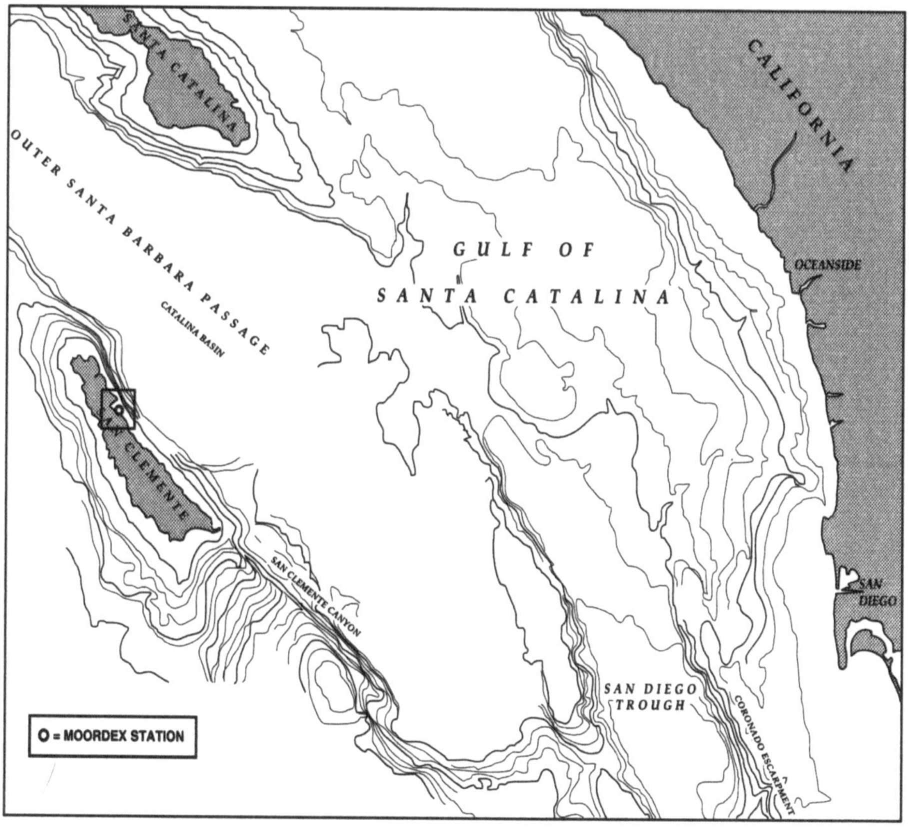

Fig. 1. Bioluminescence study area within the Gulf of Santa Catalina. Boxed area identifies NOTS pier where all plankton collections were made.

2.1.1 Laboratory measurements

Single cells in cuvettes were measured in a laboratory photometer system (Lapota et al. 1994b). This system is similar to another photometer system used in previous studies (Lapota and Losee 1984; Lapota et al. 1988, 1989, 1992) and consists of a horizontally mounted 2-inch diameter end window photomultiplier tube (PMT - RCA 8575 with an S-20 response) attached to a darkened chamber with removable opening to receive the cuvettes. Cells were stimulated to luminescence by stirring with an adjustable speed motor driving a plastic stirrer for 30 sec after which the total bioluminescence (PMT counts) was displayed. PMT counts were either logged on a notebook computer operating under Windows™ or hand recorded on data sheets. PMT dark counts were subtracted from all light output values prior to conversion to photons cell^{-1}. The system was calibrated with aliquots of the luminescent bacterium *Vibrio harveyii* measured by the Quantalum 2000 silicon-photodiode detector (Matheson et al., 1984). The detector calibration is traceable to a luminol light standard.

Following testing, the cells were individually placed in borosilicate vials and preserved in a 5% formalin solution for later microscopic identification to species level. Mean light output values for each species tested for all seasons were calculated. The Student's t test for comparison of two means was used to calculate significant differences between means of cells within species for all seasons. Critical values of the Student's t distribution were calculated using n-2 degrees of freedom.

2.2 Nutrient and chlorophyll data

Nitrate and Chl *a* levels were obtained from archived CalCOFI data bases from 1994 - 1996 for the Southern California Bight and were averaged along CalCOFI lines 90 and 93 which extend offshore west of San Diego to the north and south of SCI (Hayward et al. 1996). Nitrates (μm L^{-1}) and Chl *a* (μg L^{-1}) along each of the CalCOFI transit lines 93.26 to 93.45 and 90.28 to 90.53 were averaged from the surface to a depth of 50 m, seasonally (fall, winter, spring, summer), from spring 1994 through spring 1996. These data were used to calculate correlations with seasonal means of bioluminescence $cell^{-1}$ of the photosynthetic dinoflagellate *Pyrocystis noctiluca* and the heterotrophic dinoflagellate *Protoperidnium pellucidum*. Nutrient data was lagged by one season to calculate correlation coefficients with mean bioluminescence $cell^{-1}$. The Student t-test was used to determine significant differences among seasonal bioluminescence means for *Gonyaulax, Ceratium, Pyrocystis*, and *Protoperidinium* species.

3. Results

3.1 Bioluminescence measured with MOORDEX

A winter maximum and summer minimum in bioluminescence was measured at SCI in contrast to a maximum in the spring and minimum in the fall in San Diego Bay (SDB) (Figure 2). Mean monthly bioluminescence at SCI varied little from August 1993 - February 1996 except during a red tide in January 1995 (maximum of 2 x 10^8 photons s^{-1} ml^{-1} seawater measured in January 1995) which persisted through April (Figure 2). In contrast, seasonal changes in bioluminescence were observed in SDB. Maximum bioluminescence (1 x 10^8 photons s^{-1} ml^{-1} or greater) was measured from March through September for 1993, May through June for 1994, December through May for 1995, and March through April for 1996. Minimum values less than 1 x 10^8 photons sec^{-1} ml^{-1}) were measured in the winter (Figure 2) (Lapota et al., 1997).

3.2 Bioluminescent plankton at San Clemente Island

The numbers of bioluminescent dinoflagellates were typically lower at SCI than at SDB and ranged from 3 - 211 cells L^{-1} of seawater from August 1993 through December 1994 (Figure 3). In contrast, in SDB, maximum numbers of bioluminescent dinoflagellates were collected during the spring-summer months (2430 - 17,216 cells L^{-1}). Minimal numbers of bioluminescent dinoflagellates were collected in the winter in SDB. At SCI, the principal species observed were *G. polyedra* and several species of *Protoperidinium*. A red tide was first observed in January 1995 and persisted through April 1995. Bioluminescence during this event increased approximately 10 times above former levels for both SDB and SCI. Total and bioluminescent dinoflagellates increased to 16,727 cells L^{-1} and 15,939 cells L^{-1}, respectively in January 1995 at SCI (Figure 3). Cell numbers remained high through April 1995. At SCI bioluminescent dinoflagellates comprised a major percentage of total dinoflagellates collected (Figure 4).

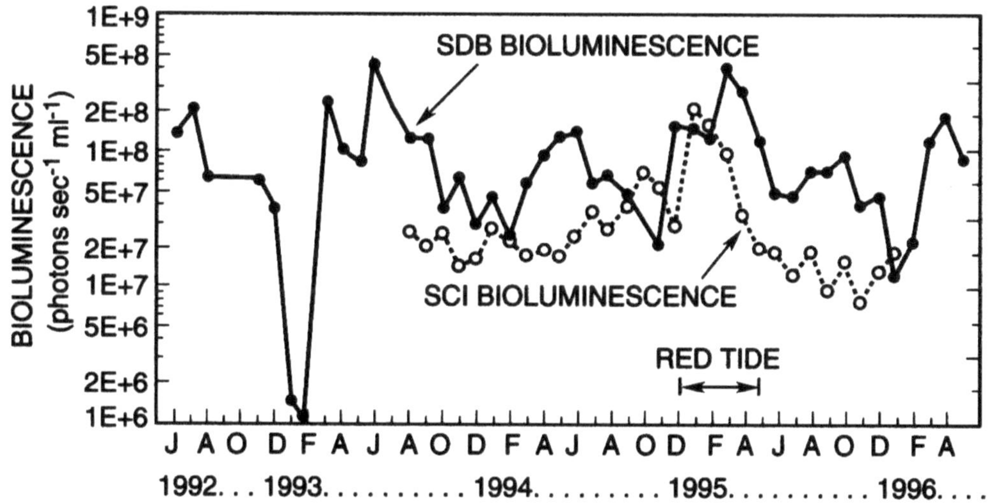

Fig. 2. Mean monthly bioluminescence trends at San Diego Bay and San Clemente Island from 1992-1996.

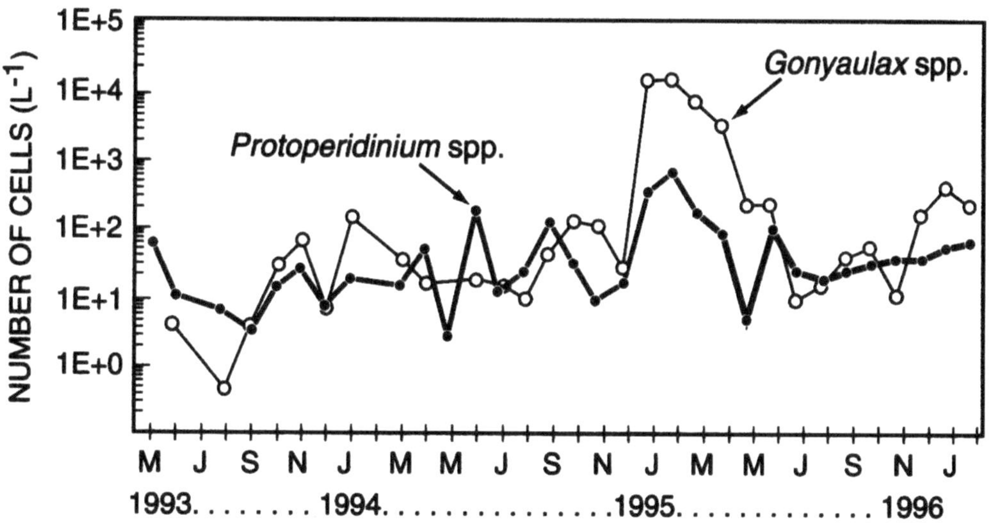

Fig. 3. The abundance of *Protoperidinium* and *Gonyaulax* dinoflagellates at San Clemente Island from 1993-1996.

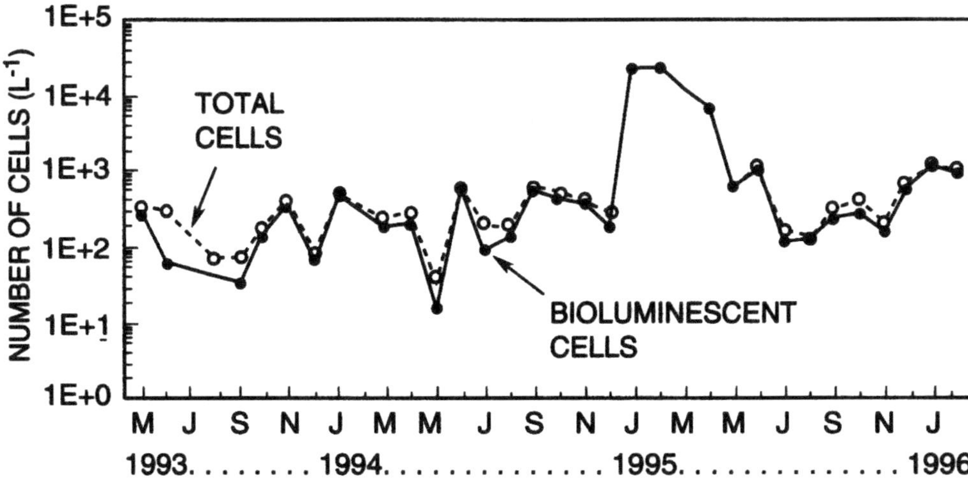

Fig. 4. Total and bioluminescent dinoflagellate cells collected monthly at San Clemente Island from 1993-1996.

3.3 Light budget analysis

The light budget analysis (number of cells L^{-1} of each bioluminescent species multiplied by the mean light output $cell^{-1}$ and then adjusted by dividing the species light contribution by the sum of all bioluminescence from all species) indicated that the *Protoperidinium* dinoflagellates produced most of the bioluminescence (Figure 5).

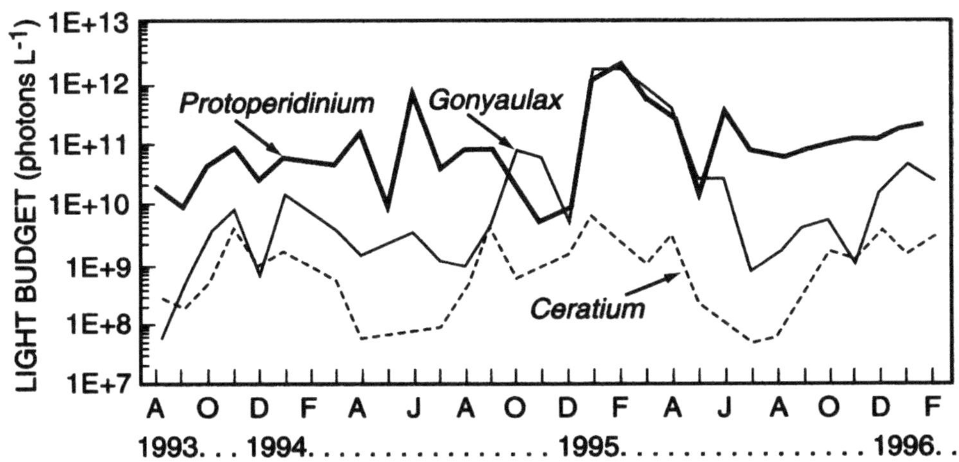

Fig. 5. Bioluminescence produced by each species (photons $liter^{-1}$) monthly at San Clemente Island from 1993-1996.

Gonyaulax spp. and *Ceratium fusus* contributed less. *Protoperidinium* contributed more than 80% of all bioluminescence for 60% of all months (30 months) and more than 50% of all bioluminescence for 77% of all months. In contrast, *Gonyaulax* contributed 80% of all bioluminescence for just 1 month (3.3% of all months) and 50% of all bioluminescence for only 10% of all months. During the red tide in the winter and spring of 1995, *Gonyaulax* contributed 59%, 42%, 58%, 48%, and 27% of all bioluminescence for the months of January through May 1995, respectively (Figure 6). The open ocean bioluminescent dinoflagellate *Pyrocystis noctiluca* was also found in monthly samples. *Protoperidinium curtipes, P. depressum, P. divergens, P. leonis,* and *P. steinii* were present in the spring and summer months while *Gonyaulax* became more prevalent in the fall and winter months.

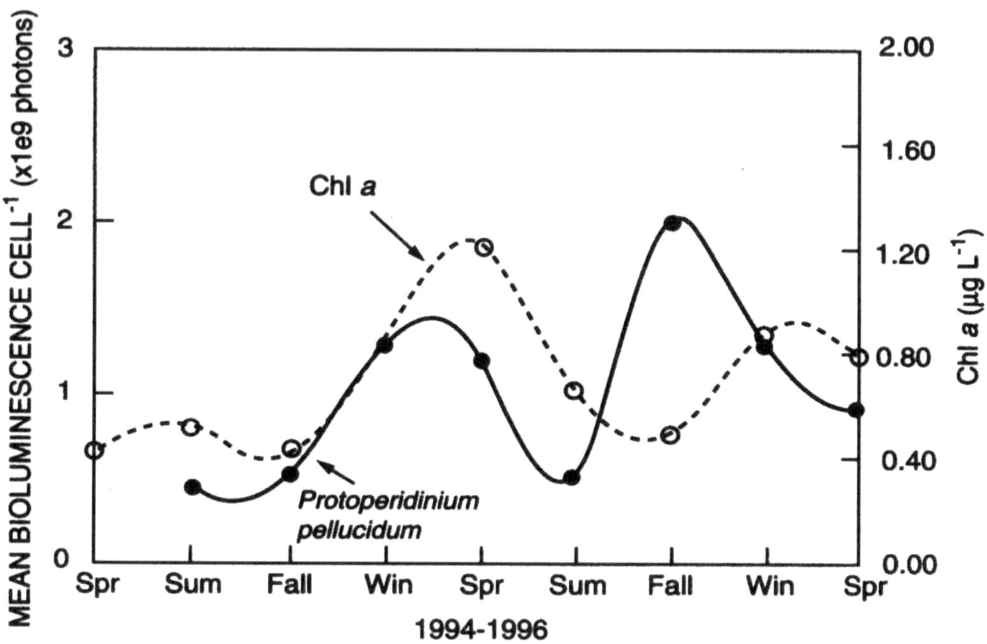

Fig. 6. (a) Seasonal differences in *Protoperidinium pellucidum* mean bioluminescence and mean Chl *a* for waters surrounding San Clemente Island 1994-1996.

3.4 Light output from Dinoflagellates

Three to 6 recurrent species of dinoflagellates were usually found in the collected plankton samples for each season. In 5 of the 9 seasons, more than 45 cells isolated proved to be luminescent and were identified to species level. In all, bioluminescence from 469 identified cells representing 5 genera and 13 species were measured (Table 1). Cells of *Gonyaulax polyedra, Ceratium fusus, Pyrocystis noctiluca,* and the heterotroph *Protoperidinium pellucidum* were found more frequently in plankton collections than other species, and from these data, seasonal mean bioluminescence cell^{-1} was determined.

Mean light output cell^{-1} for all dinoflagellates is listed in Tables 2,3,4,5 and in Figures 6-9. The mean light output for all quarters for all species is listed in Table 2 and ranges from 1.6e8 photons cell^{-1} for *Ceratium fusus* to 1.8e10 photons cell^{-1} for *Pyrocystis noctiluca. G. grindleyi* was the brightest species of *Gonyaulax* tested and was approximately 3 times brighter than *G. polyedra* (Table 2). Cells of species of *Protoperidinium* ranged from 1 to 4e9 photons cell^{-1} (Tables 2, 3; Figure 6a).

Species	Season Tested
G. grindley	Fall'94, Win'95
G. polyedra	Win'95, Spr'95; Fall'95, Win'96
G. polygramma	Spr'94, Fall'95, Win'96
G. spinifera	Sum'95, Fall'95
Ceratium fusus (small cells)	Fall'94, Win'95; Spr'95, Fall'95; Win'96, Spr'96
C. fusus (large cells)	Fall'95, Win'96; Spr'96
Protoperidinium curtipes	Spr'94, Sum'94; Spr'96
P. depressum	Spr'94, Sum'94; Win'96
P. divergens	Spr'94, Sum'94
P. leonis	Spr'94, Spr'95
P. pellucidum	Sum'94, Fall'94; Win'95, Spr'95; Sum'95, Fall'95; Win'96, Spr'96
P. steinii	Fall'94
Noctiluca miliaris	Spr'95
Pyrocystis noctiluca	Spr'94, Sum'94 ; Fall'94, Win'95 ; Sum'95, Fall'95; Win'96, Spr'96

Table 1. Bioluminescent dinoflagellate species surveyed in study.

Species	Mean light output (photons cell^{-1})	Standard Deviation	Number of cells tested
***Ceratium fusus* (small cells)**	1.5e8	1.9e8	52
Gonyaulax polyedra	2.4e8	2.2e8	56
Gonyaulax spinifera	6.1e8	4.5e8	19
Gonyaulax polygramma	6.3e8	4.2e8	36
***Ceratium fusus* (large cells)**	8e8	5.7e8	26
Gonyaulax grindleyi	8.5e8	5.2e8	10
Protoperidinium steinii	1e9	1.1e9	5
Protoperidinium pellucidum	1.2e9	1e9	112
Protoperidinium divergens	2.8e9	3.9e9	19
Protoperidinium depressum	3.1e9	2.8e9	18
Protoperidinium curtipes	4e9	4.7e9	17
Protoperidinium leonis	4.4e9	3e9	7
Pyrocystis noctiluca	1.8e10	9.7e9	105

Table 2. Comparative light output for all species 1994-1996. Listing is by mean light output.

Species	Spr'94	Sum'94	Fall'94	Win'95	Spr'95	Sum'95	Fall'95	Win' 96	Spr'96
P. de.	1.6e9; 7 cells	4.1e9; 9 cells						3.6e9; 2 cells	
P. di.	4.4e9; 8 cells	1.6e9; 11 cells							
P. cur.	5.9e9; 9 cells	7.9e8; 3 cells							2.5e9; 5 cells
P. leo	6.3e9; 4 cells				1.8e9; 3 cells				
P. stei.			4.3e8; 3 cells						1.9e9; 3 cells

Table 3. Light output for other *Protoperidinium* spp. (mean photons cell^{-1}) 1994 – 1996.

Species	Spr'94	Sum'94	Fall'94	Win'95	Spr'95	Sum'95	Fall'95	Win'96	Spr'96
G. poly.	6.2e8;9 cells						5.2e8; 12 cells	7.3e8; 15 cells	
G. spin.						2.2e8; 7 cells	8.4e8; 12 cells		
G. grind.			8.9e8; 8 cells	6.8e8; 2 cells					
G. polyed.				9.8e7 ; 19 cells	1.1e8; 12 cells		4.2e8; 14 cells	4.2e8; 11 cells	

Table 4. Light output for other *Gonyaulax* species (mean photons cell-1) 1994 – 1996.

Cells	Spr'94	Sum'94	Fall'94	Win'95	Spr'95	Sum'95	Fall'95	Win'96	Spr'96
Small cells			3.6e8; 9 cells	5.7e7; 7 cells	2.1e7; 1 cell		8.4e7; 13 cells	1.6e8; 19 cells	2.3e8; 3 cells
Large cells							8.4e8; 6 cells	6.3e8; 15 cells	1.3e9; 5 cells

Table 5. Light output for *Ceratium fusus* (in mean photons cell^{-1})1994-1996.

Win'95 bioluminescence is significantly greater than	**Sum'94**	**Fall'94**	**Spr'95**
	p< 0.10	p < 0.001	n.s.
Fall'95 bioluminescence is significantly greater than	**Sum'95**	**Win'96**	**Spr'96**
	p < 0.01	n.s.	p < 0.01

*levels of significance for two-tail student t test

Table 6. Significant differences of light output seasonally in *Protoperidinium pellucidum* from 1994 - 1996*.

Seasons between years

Sum'94 - Sum'95 bioluminescence	**n.s.**
Fall'94 - Fall'95 bioluminescence	Fall'95 > Fall'94; $p < 0.001$
Win'95 - Win'96 bioluminescence	n.s.
Spr'95 - Spr'96 bioluminescence	n.s.

n.s. = not significant

3.4.1 *Pyrocystis noctiluca*

Significant differences in mean bioluminescence cell^{-1} were observed between spring '94 and summer '94, between spring '94 and fall '94, between spring '94 and winter '95, and between spring '94 and spring '96 (Table 7). Spring '96 cells produced more bioluminescence than spring '94 cells as did winter '96 cells when compared to winter '95 cells (Table 7). *Gonyaulax* spp.

	Spr'94	Sum'94	Fall'94	Win'95	Spr'95	Sum'95	Fall'95	Win'96	Spr'96
Sum'94	p<0.3								
Fall'94	p<0.2	P<0.4*		P<0.6*					
Win'95	p<0.3	P<0.9*							
Spr'95		P<0.6*							
Sum'95		P<0.6*						P<0.8*	P<0.7*
Fall'95			P<0.5*			P<0.8*	P<0.3	P<0.7*	
Win'96				P<0.2					P<0.5*
Spr'96	p<0.1								

Table 7. Significant differences of light output seasonally in *Pyrocystis noctiluca* from 1994 - 1996 (* = not significant).

Seasons between years Fall'94 - Fall'95

Spr'94 - Spr'96	**Spr'96 > Spr'94**	$p < 0.1$
Sum'94 - Sum'95	Sum'95 > Sum'94	p< 0.6*
Fall'94 - Fall'95	Fall'95 > Fall'94	p< 0.5*
Win'95 - Win'96	Win'96 > Win'95	$p < 0.2$

*= not significant

Significant differences in mean cell light output were measured among *Gonyaulax polyedra, G. spinifera*, and *G. polygramma* (Table 8). Cells of *G. polyedra* of fall '95 produced significantly more light than cells tested previously in the winter and spring '95 ($p < 0.001$; Table 8). Cells tested in the winter '96 were significantly brighter than the previous spring '95. Yearly differences were also observed between both winters of '95 and '96 where '96 cells > '95 cells. The fall '95 cells of *G. spinifera* exhibited more bioluminescence than in the summer while winter '96 *G. polygramma* cells were measurably brighter than earlier in the fall (Table 8).

	G. polyedra	
Win'95 - Spr'95	P<0.8	n.s.
Fall'95 – Win'95	Fall > Win	$p < 0.001$
Fall '95 – Spr'95	Fall > Spr	$p < 0.001$
Win'96 – Win'95	Win'96 > Win'95	$p < 0.001$
Win'96 – Spr'95	Win > Spr	$p < 0.001$
Win'96 – Fall'95	$P < 0.9$	n.s.
	G. spinifera	
Fall'95 – Sum'95	Fall > Sum	$P < 0.001$
	G. polygramma	
Win'96 – Fall'95	Win > Fall	$P < 0.2$

Table 8. Significant differences of light output seasonally in *Gonyaulax* species from 1995 - 1996.

3.4.2 *Protoperidinium*

The winter '95 *Protoperidinium pellucidum* cells were observed to produce more bioluminescence than cells from summer '94 and fall '94 (Table 6). They were not significantly different than cells tested in spring '95. Cells of *P. pellucidum* in the fall '95 were significantly brighter than cells tested in summer '95, and spring '96, but not winter '96 (Table 6). A significant seasonal difference between years was observed for the fall quarters. These cells were also significantly brighter in Fall '95 than in the fall '94 (Table 6). Significant differences in light output were not found for summer, winter, and spring between 1994 and 1995.

Limited measurements were conducted on *P. curtipes* and *P. divergens*. These cells produced more bioluminescence in the spring '94 than the following summer while *P. depressum* cells in the summer '94 were brighter than the previous spring months (Table 9). *P. leonis* cells were observed to produce more bioluminescence in spring '94 than in spring '95.

P. curtipes	**Spr'94 - Sum'94**	**Spr > Sum**	**$p < 0.2$**
P. depressum	Sum'94 - Spr'94	Sum > Spr	$p < 0.1$
P. divergens	Spr'94 - Sum'94	Spr > Sum	$p < 0.2$
P. leonis	Spr'94 - Spr'95	Spr'94 > Spr'95	$p < 0.05$

Table 9. Significant differences of light output seasonally in *Protoperidinium* species from 1994 - 1995.

3.4.3 *Ceratium fusus*

Significant seasonal differences of light output were observed from 1994-1996 in small and larger cells of *C. fusus*. Fall '94 cells produced more bioluminescence than winter '95 cells and winter '96 cells. In contrast, winter '96 cells produced more bioluminescence than fall '95 cells (Table 10). Between years, fall '94 cells were brighter than fall '95 cells while winter '96 cells exhibited more bioluminescence than winter '95 cells. Light output in larger cells of *C. fusus* showed that spring '96 were brighter than in fall '95 and winter '96. Larger cells of *C. fusus* were significantly brighter than smaller *C. fusus* cells for fall '95, winter '96 and spring '96.

	SMALL CELLS	
Fall'94 - Win'95	**Fall > Win**	**$p < 0.02$**
Fall'94 - Win'96	Fall > Win	**$p < 0.05$**
Fall'94 - Spr'96	$p < 0.5$	**n.s.**
Win'96 - Fall'95	Win > Fall	**$p < 0.2$**
	Seasons between years	
Fall'94 - Fall'95	Fall'94 > Fall'95	**$p < 0.01$**
Win'96 - Win'95	Win'96 > Win'95	**$p < 0.1$**
	LARGE CELLS	
Spr'96 - Win'96	Spr > Win	**$p < 0.05$**
Spr'96 - Fall'95	Spr > Fall	**$p < 0.2$**
Fall'95 - Win'96	$p < 0.5$	**n.s.**
	Difference in light output between large and small cells	
Fall'95	Large cells > small cells	**$p < 0.001$**
Win'96	Large cells > small cells	**$p < 0.01$**
Spr'96	Large cells > small cells	**$p < 0.05$**

n.s. = not significant

Table 10. Significant differences of light output seasonally in *Ceratium fusus* from 1994 – 1996.

3.4.4 Seasonal trends of dinoflagellate bioluminescence, nitrates, and Chl *a*

Maximum bioluminescence in *Protoperidinium pellucidum, Pyrocystis noctiluca, Ceratium fusus,* and *Gonyaulax polyedra* was observed in either fall or winter for 1994–1996. Maximum bioluminescence cell^{-1} in *P. pellucidum* occurred in winter 1995 and during the following fall. Bioluminescence in this species was lower in summer '94 and again in summer '95 (Figures 6a,b). The percent of peak bioluminescence (the peak refers to maximum bioluminescence measured in winter '95 and fall '95) varies with season with values as low as 30% for summer '94 and 20% for summer '95 (Figure 6b). Because *Protoperidinium* dinoflagellates are nutritionally dependent on other algal species, any change of algal biomass in waters adjacent to SCI might impact the growth, maintenance and luminescent potential of *Protoperidinium*. Extracted Chl *a* showed a similar periodicity as did mean bioluminescence cell^{-1} in *P. pellucidum* from summer '94 through summer '95. Maximum Chl *a* and mean bioluminescence cell^{-1} levels increased in winter 1995 and in spring 1995 (Figure 7a). The seasonal mean bioluminescence cell^{-1} and the seasonal mean Chl *a* in the waters off SCI were positively correlated (Figure 6c). Bioluminescence in *P. pellucidum* was not significantly different between winter and spring quarter 1995, but was significantly brighter than the preceding summer and fall quarters and the ensuing summer 1995 quarter (Table 6).

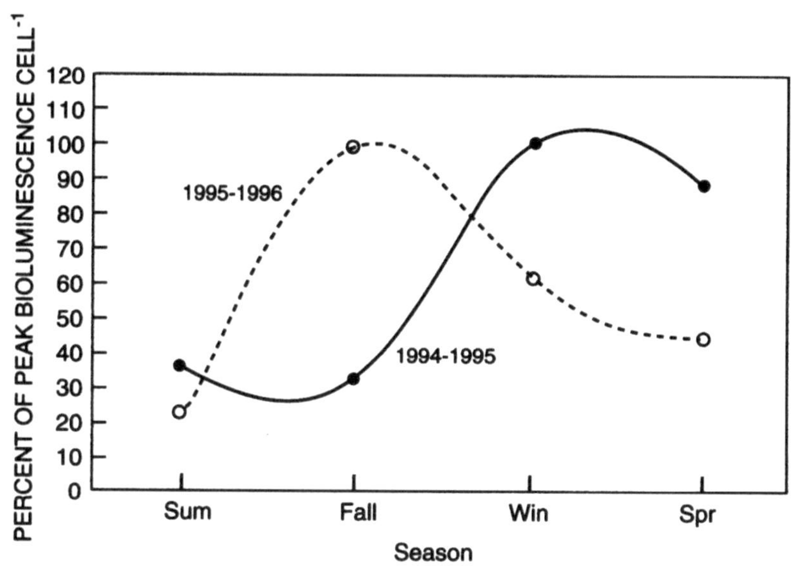

Fig. 6. (b) 1994 to 1996 year comparison of percent of maximum bioluminescence of *Protoperidinium pellucidum* at San Clemente Island.

Maximum bioluminescence in *Pyrocystis noctiluca* was measured in the fall of 1994 and 1995 while minimum bioluminescence was measured in the spring of 1994 and 1996 (Figures 7a-d). Maximum levels of nitrates for both years were in the summer, a time of maximum upwelling in the Southern California Bight. Maximum levels of nitrates preceded maximum bioluminescence for both years (Figure 7c) to possibly explain these seasonal changes in a photosynthetic bioluminescent dinoflagellate. Nitrate levels (μM L^{-1}) were averaged seasonally and compared with bioluminescence (Figure 7b). Both trends exhibit similar amplitudes, i.e., lower levels of nitrates were present in summer and fall of 1994 when compared to summer and fall nitrate levels in1995. Bioluminescence showed similar trends for the same period (r= 0.859; p < 0.02). Less bioluminescence was measured in cells collected in fall '94 than in cells measured in fall '95. Spring bioluminescence was less than fall bioluminescence in *P. noctiluca* for both years (Figure 8d). Spring '94 bioluminescence was approximately 40% of the maximum of fall '94 and approximately 65% of maximum for fall '95.

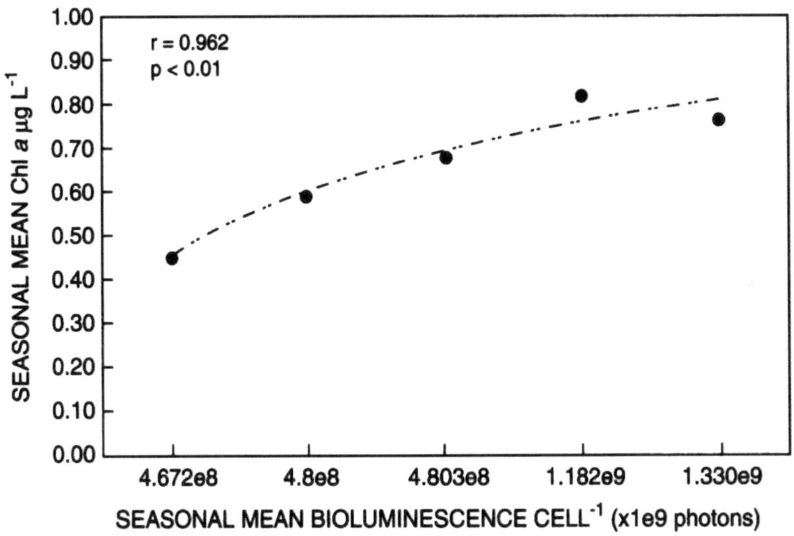

Fig. 6. (c) Correlation of seasonal mean bioluminescence in *Protoperidinium pellucidum* with seasonal mean Chl *a* at san Clemente Island from summer 1994 through sumer 1995.

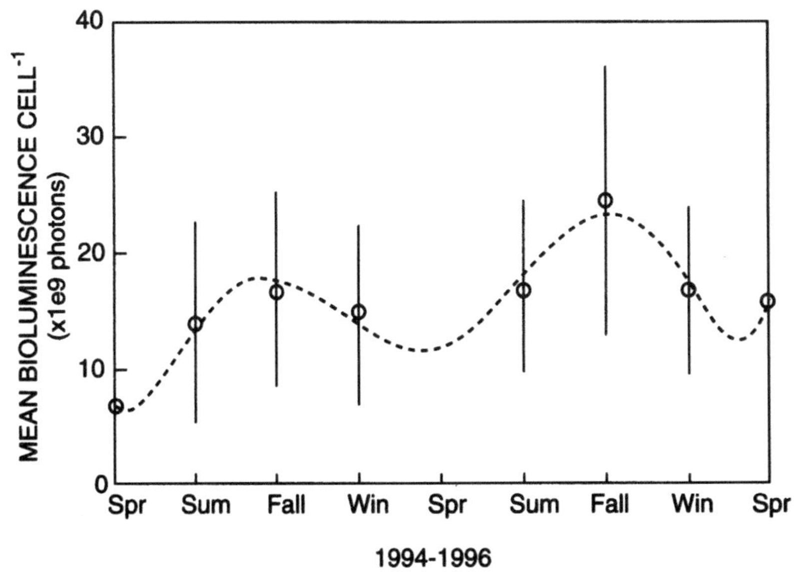

Fig. 7. (a) Seasonal differences in *Pyrocystis noctiluca* bioluminescence at San Clemente Island from 1994-1996. Error bars represent 1 standard deviation of the seasonal means.

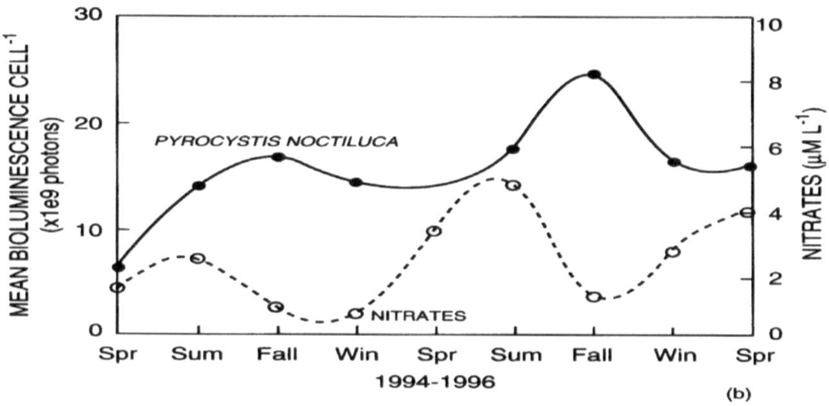

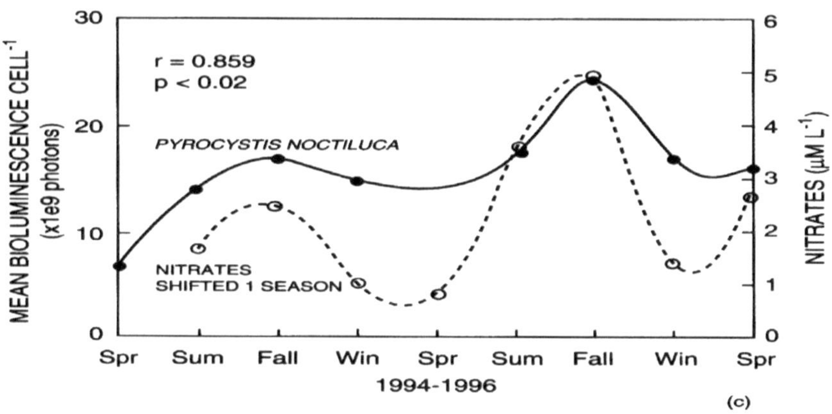

Fig. 7. (b) Seasonal differences in mean bioluminescence of *Pyrocystis noctiluca* and mean nitrates in waters surrounding San Clemente Island from 1994-1996. (c) Correlation of seasonal means of *Pyrocystis noctiluca* bioluminescence and mean nitrates. Mean nitrates were shifted to the right by 1 season to illustrate temporal and magnitude similarities in both trends for both years.

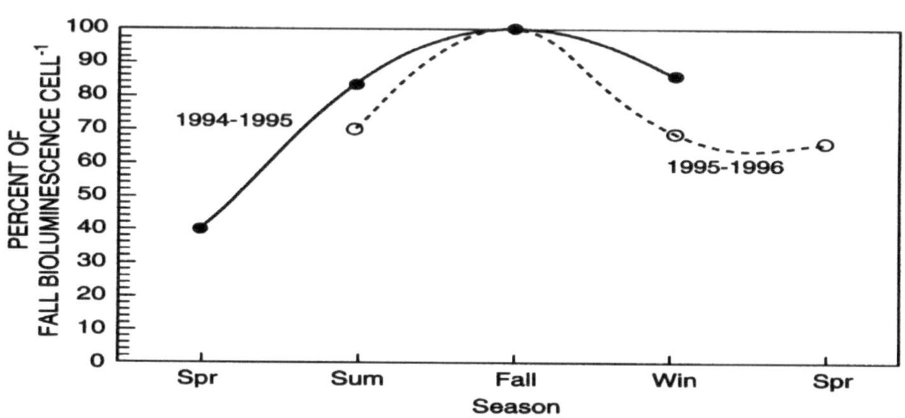

Fig. 7. (d) 1994 to 1996 comparison of percent of maximum bioluminescence of *Pyrocystis noctiluca* at San Clemente Island.

Seasonal and yearly differences of bioluminescence in *Gonyaulax polyedra* and *Ceratium fusus* were also observed (Table 8, 10). While both data sets are incomplete with respect to a continuous record, the data do show a maximum bioluminescence in fall '94 for *C. fusus* and differences in bioluminescence in *G. polyedra* between winter '95 and winter '96.

4. Discussion

The data agree that dinoflagellate bioluminescence has a marked seasonality in the open ocean which is affected by regional environmental events such as upwelling and rainstorms resulting in enhanced terrestrial runoff. Other laboratory observations support the view that nutritional requirements are important in determining bioluminescent capacity (Sweeney 1971). For example, *Protoperidinium* dinoflagellates underwent increases in bioluminescence potential when fed to excess with diatoms (Buskey et al., 1992; Latz 1993) In an earlier study, when the heterotroph *Noctiluca miliaris* was fed with the flagellate *Dunaliella* more light was emitted than from unfed *Noctiluca* (Sweeney 1981). It was also reported that a strain of this dinoflagellate carrying a photosynthetic algal symbiont produced bioluminescence which was proportional to the light intensity at which the symbiont was grown, suggesting a nutritive contribution by the algal symbiont (Sweeney 1981; Sullivan and Swift 1995). Certainly other observations support the view that nutritional requirements are important in determining bioluminescent capacity (Sweeney 1971). Laboratory investigations have also shown that increased irradiance elevates photosynthesis with consequent increased bioluminescence (Sweeney et al., 1959; Sweeney 1981; Swift et al., 1981; Sullivan and Swift 1995).

When cultures of *Gonyaulax polyedra* were maintained in artificial seawater media for periods of up to 37 days, mean bioluminescence decreased by almost a factor of 10 when compared to cells after 5-13 days in culture (Sweeney 1981). Because cell numbers increased throughout the study (from 6,300 cells ml^{-1} to 21,830 cells ml^{-1}), auto-toxicity is not a likely explanation of this significant decrease in bioluminescence, leaving nutrient limitation a possibility. This was tested by Sweeney in the same report with the finding that nitrates and phosphates appeared to enhance cell growth, but not bioluminescence capacity, while only trace levels of iron sequestrine supported maximum bioluminescence and cell growth. At higher concentrations, iron sequestrine appeared to reduce the bioluminescent capacity and cell division in *G. polyedra* (Sweeney 1981). The bioavailability of nutrients and trace metals are often not addressed with respect to impact on different physiological mechanisms (cell division *vs* bioluminescent capacity) within the same cell. Particularly in coastal waters environmental contaminants might complicate interpretation of nutrient effects. Thus bioluminescence enhancement has been observed in toxicity studies using *G. polyedra* incubated for up to 4 days with bay sediment pore waters. Ammonia is commonly found in sediment pore waters and levels of 200-400 $\mu g\ L^{-1}$ have been observed to increase light output 3-4 times above controls. The data suggest that the organism may be responding to a readily available increased source of nitrogen, possibly an example of hormesis (Unpublished data, Lapota and Liu 1997).

In the present study, we have observed seasonal trends in nitrates, Chl *a*, and bioluminescence in numerous species of bioluminescent dinoflagellates. Maximum bioluminescence in *Protoperidinium pellucidum* was observed in winter '95 and in fall '96 which might be explained by the availability in the diet of diatoms and *Gonyaulax polyedra* (Figure 7a).

Increased levels of Chl *a* were measured in the winter and spring '95 and were strongly correlated with increased *P. pellucidum* bioluminescence. Species of *Protoperidinium* are known to graze on *G. polyedra* in laboratory studies (Buskey et al., 1992; Latz and Jeong 1993; Jeong and Latz 1994). Latz (1993) demonstrated the maintenance of *Protoperidinium divergens* growth, survival, and bioluminescence capacity when grazing on a variety of dinoflagellates, but found maximum bioluminescence when the diet was solely *G. polyedra*. The winter '95 period within the Southern California Bight was characterized by an extensive red tide and extended from Santa Barbara south to San Diego and west to San Clemente Island. *G. polyedra* was the principal dinoflagellate present, reaching concentrations of approximately 16,000 cells l^{-1} in January 1995, although increases in *Protoperidinium* spp. were also observed (Lapota et al., 1997). Heavy rainfall was recorded during this winter period (17-18 inches, as compared with the norm of 5-10 inches) and consequently extensive runoff was observed along the entire southern California coast. Total bioluminescence (photons ml^{-1} $year^{-1}$) was positively correlated with rainfall (inches $year^{-1}$) for a 4 year period in San Diego Bay (1992-1996) (Lapota et al., 1997). Nitrates and trace metals are carried off from land with the runoff into surface waters (Dugdale and Goering 1967). Thus, the *G. polyedra* red tide was probably triggered by extensive runoff including nutrients such as nitrates from this "wet" year which in turn stimulated growth of phytoplankton grazed by *Protoperidinium pellucidum* and other *Protoperidinium* species. Increases in bioluminescence were observed in more than 60% of all rain events from 1992-1996 in San Diego Bay (Lapota et al., 1997). Others have observed these sudden blooms and they often occur in spring or summer following heavy rains that produce nutrient-rich land runoff (Eppley, 1986). The reason for a fall '95 peak in *P. pellucidum* bioluminescence is unknown but may be due to grazing by *P. pellucidum* on lower numbers of *G. polyedra* and other algal cells. Upwelling and nitrate levels (Figure 8b) were greatest during the summer months and could result in a later increase in photosynthetic biomass in the fall. However, Chl *a* levels were actually low during the period when bioluminescence was high and may indicate previous grazing by *P. pellucidum*. Seasonal mean Chl *a* and mean bioluminescence $cell^{-1}$ were strongly correlated ($r = 0.962$; $p < 0.02$) for 1994-1995 which may suggest that as Chl *a* levels increased, so did bioluminescence $cell^{-1}$ (Figure 7c). These field measurements support previous laboratory studies (Buskey et al., 1992; Latz and Jeong 1993).

Both nitrates and mean bioluminescence $cell^{-1}$ in *Pyrocystis noctiluca* show similar trends temporally and in magnitude (Figures 7b, 7c). Peak levels of nitrates were found in the summer months followed by increases in bioluminescence during the fall months. Nitrate levels were greater in summer '95 than in summer '94. Peak bioluminescence was also greater in fall '95 than in fall '94. Lagging data comparisons + 1 season for nitrates produced a strong correlation with bioluminescence for the entire 2 year period ($r = 0.859$; $p < 0.02$). That is, lower nitrate levels measured in 1994 correlated with lower bioluminescence in 1994 while greater nitrate levels correlated with an increased bioluminescence (Figure 7c). Peak bioluminescence was also observed to occur in the fall for both years (Figure 7d). The effect of nitrate on bioluminescent capacity within photosynthetic dinoflagellates is unclear, but perhaps may be related to the overall health of the cell and how *Pyrocystis*, *Gonyaulax* spp. and *Ceratium* partition their metabolic resources when nitrate satiated. Others have observed the photosynthesis-irradiance relation to bioluminescence capacity. That is, cells grown at higher irradiance levels produce more photosynthetic products which may be diverted to the bioluminescence system (Sweeney et al., 1959; Sweeney 1981; Swift et al., 1981; Sullivan and Swift 1995). It is very possible that increased levels of nutrients from upwelling and storm runoff events may override diminished irradiance levels found during the fall and winter months to explain maximum bioluminescence observed in these species.

5. Conclusion

A significant portion of bioluminescence in all oceans is produced by dinoflagellates. The number of bioluminescent species and their relative abundance changes temporally and spatially. There is evidence that dinoflagellates exhibit changes in per cell bioluminescence magnitude which may be attributable to environmental conditions such as light, temperature, and nutrient history. In the present study, photosynthetic and heterotrophic dinoflagellates were collected and tested for bioluminescence on a quarterly basis from 1994-1996 at San Clemente Island, located 100 km off the Southern California coast. Per cell bioluminescence was measured for the phototrophs *Ceratium fusus, Pyrocystis noctiluca, Gonyaulax polyedra* as well as 3 other species of *Gonyaulax*, and in 6 species of the heterotroph *Protoperidinium*. Our data strongly suggests that dinoflagellates have a marked seasonality in the open ocean which may be attributable to regional environmental events such as upwelling and associated winter storm land runoff. We observed correlations between surface (0-50m) nitrates and Chl *a* with bioluminescence in *Pyrocystis noctiluca* and *Protoperidinium pellucidum*. Increased levels of Chl *a* measured in the winter and spring '95 correlated with increased *P. pellucidum* bioluminescence. Both nitrates and mean bioluminescence cell^{-1} in *P. noctiluca* showed similar trends temporally and in magnitude. Peak levels of nitrates were found in the Southern California Bight in the summer months followed by increases in bioluminescence during the fall months. Peak bioluminescence was observed to occur in the fall for both years in *P. noctiluca* whereas peak bioluminescence in *P. pellucidum* was measured in winter '95 and later in fall '95.

6. Acknowledgments

We gratefully acknowledge financial support from the Office of Naval Research, Code 322BC, Arlington, VA and the Naval Space and Warfare Systems Center, San Diego, CA for conducting these studies. Special thanks are due to Dr. Jim Case (University of California, Santa Barbara) for his continuing guidance in this study and Ms. Connie H. Liu (Naval Space and Warfare Systems Center, San Diego, CA) for providing mean Chl *a* and nitrate values for this study.

7. References

Batchelder, H.P., Swift, E. (1989). Estimated near-surface mesoplanktonic bioluminescence in the western North Atlantic during July 1986. Limnol. Oceanogr. 34: 113-128

Batchelder, H.P., Swift, E., Van Keuren, J.R. (1990). Pattern of planktonic bioluminescence in the northern Sargasso Sea: seasonal and vertical distribution. Mar. Biol. 104: 153-164

Batchelder, H.P., Swift, E., Van Keuren, J.R. (1992). Diel patterns of planktonic bioluminescence in the northern Sargasso Sea. Mar. Biol. 113: 329-339

Bityukov, E.P., Rybasov, V.P., Shaida, V.G. (1967). Annual variations of the bioluminescent field intensity in the neritic zone of the Black Sea. Oceanology 7 (6): 848-856

Buskey, E.J., Strom, S., Coulter, C. (1992). Bioluminescence of heterotrophic dinoflagellates from Texas coastal waters. J. exp. mar. Biol. Ecol. 159: 37-49

Colebrook, J.M., Robinson, G.A. (1965). Continuous plankton records: seasonal cycles of phytoplankton and copepods in the north-eastern Atlantic and the North Sea. Bull. Mar. Ecol. 6: 123-139

Dodge, J.D., Hart-Jones, B. (1977). The vertical and seasonal distribution of dinoflagellates in the North Sea, II., Blyth 1973-1974 and Whitby 1975. Bot. Mar. 20: 307-311

Dodge, J.D. (1989). Some revisions of the family Gonyaulacaceae (Dinophyceae) based on a scanning electron microscope study. Bot. Mar. 32: 275-298

Dugdale, R.C., Goering, J.J. (1967). Uptake of new and regenerated forms of nitrogen and primary productivity. Limnol. Oceanogr. 12: 196-206

Eppley, R.W. (1986). People and the plankton. In: Eppley, R.W. (ed.) Plankton Dynamics of the Southern California Bight, Springer-Verlag, New York, Inc. pp. 289-303

Hayward, T.L., Cummings, S.L., Cayan, D.R., Chavez, F.P., Lynn, R.J., Mantyla, A.W., Niiler, P.P., Scwing, F.B., Veit, R.R., Venrick, E.L. (1996). The state of the California Current in 1995: Continuing declines in macrozooplankton biomass during a period of nearly normal circulation. CALCOFI Reports 37: 22-37

Jeong, H.J., Latz, M.I. (1994). Growth and grazing rates of the heterotrophic dinoflagellates *Protoperidinium* spp. on red tide dinoflagellates. Mar. Ecol. Prog. Ser. 106:173-185

Lapota, D., Losee, J.R. (1984). Observations of bioluminescence in marine plankton from the Sea of Cortez. J. exp. mar. Biol. Ecol. 77: 209-240

Lapota, D., Galt, C., Losee, J.R., Huddell, H.D., Orzech, K., Nealson, K.H. (1988). Observations and measurements of planktonic bioluminescence in and around a milky sea. J. exp. mar. Biol. Ecol. 119: 55-81

Lapota, D., Geiger, M.L., Stiffey, A.V., Rosenberger, D.E., Young, D.K. (1989). Correlations of planktonic bioluminescence with other oceanographic parameters from a Norwegian fjord. Mar. Ecol. Prog. Ser. 55: 217-227

Lapota, D., Young, D.K., Bernstein, S.A., Geiger, M.L., Huddell, H.D., Case, J.F. (1992a). Diel bioluminescence in heterotrophic and photosynthetic marine dinoflagellates in an Arctic Fjord. J. mar. biol. Ass. U.K. 72: 733-744

Lapota, D., Rosenberger, D.E., Lieberman, S.H. (1992b). Planktonic bioluminescence in the pack ice and the marginal ice zone of the Beaufort Sea. Mar. Biol. 112: 665-675

Lapota, D., Paden, S., Duckworth, D., Rosenberger, D.E., Case, J.F. (1994a). Coastal and oceanic bioluminescence trends in the southern California bight using MOORDEX bathyphotometers. In: Campbell, A.K., Kricka, L.J., Stanley, P.E. (eds.). Bioluminescence and Chemiluminescence. John Wiley & Sons, Chicester, England, p 127-130

Lapota, D., Rosenberger, D.E., Duckworth, D. (1994b). A bioluminescent dinoflagellate assay for detecting toxicity in coastal waters. In: Campbell, A.K., Kricka, L.J., Stanley, P.E. (eds.). Bioluminescence and Chemiluminescence. John Wiley & Sons, Chicester, England, p 156-159

Lapota, D., Duckworth, D., Groves, J., Rosen, G., Rosenberger, D., Case, J.F. (1997). Long term dinoflagellate bioluminescence, chlorophyll, and their environmental correlates in southern California coastal waters (submitted)

Latz, M.I., Jeong, H.J. (1993). Effect of dinoflagellate diet and starvation on the bioluminescence of the heterotrophic dinoflagellate, *Protoperidinium divergens*. ONR Bioluminescence Symposium, November 1993, Maui, abstract p. 61

Matheson, I.B.C., Lee, J., Zalewski, E.F. (1984). A calibration technique for photometers. Ocean Optics 7 (489): 380-381

Seliger, H.H., Biggley, W.H., Swift, E. (1969). Absolute values of photon emission from the marine dinoflagellates *Pyrodinium bahamense, Gonyaulax polyedra, and Pyrocystis lunula*. Photochem. Photobiol. 10: 232-277

Seliger, H.H., Biggley, W.H. (1982). Optimization of bioluminescence in marine dinoflagellates. Paper presented at Annual Meeting, American Society of Limnology and Oceanography, AGU, San Francisco, CA, December.

Seliger, H.H., Fastie, W.G., Taylor, W.R., McElroy, W.D. (1961). Bioluminescence in Chesapeake Bay. Science 133: 699-700

Sullivan, J.M., Swift, E. (1995). Photoenhancement of bioluminescence capacity in natural and laboratory populations of the autotrophic dinoflagellate *Ceratium fusus* (Ehrenb.) Dujardin. J. Geophy. Res. 100 (C4): 6565-6574

Sweeney, B.M., Haxo, F.T., Hastings, J.W. (1959). Action spectra for two effects of light on luminescence in *Gonyaulax polyedra.* J. Gen. Physiol. 43: 285-299

Sweeney, B.M. (1971). Laboratory studies of a green *Noctiluca* from New Guinea. J. Phycol. 7: 53-58

Sweeney, B.M. (1981). Variations in the bioluminescence per cell in dinoflagellates. In: Nealson, K.H.(ed.)., Bioluminescence Current Perspectives. Burgess Publishing, Minneapolis, p 90-94

Swift, E., Meunier, V.A., Biggley, W.H., Hoarau, J., Barras, H. (1981). Factors affecting bioluminescent capacity in oceanic dinoflagellates. In: Nealson, K.H. (ed.)., Bioluminescence Current Perspectives. Burgess Publishing, Minneapolis, p 95-106

Swift, E., Sullivan, J.M., Batchelder, H.P., Van Keuren, J., Vaillancourt, R.D. (1995). Bioluminescent organisms and bioluminescent measurements in the North Atlantic Ocean near latitude 59.5°N, longitude 21°W. J. Geophy. Res. 100 (C4): 6527-6547

Tett, P.B. (1971). The relation between dinoflagellates and the bioluminescence of sea water. J. mar. biol. Ass. U.K. 51: 183-206

Tett, P.B., Kelly, M.G. (1973). Marine bioluminescence. Oceanogr. Mar. Biol. Annu. Rev. 11: 89-173

Yentsch, C.S., Laird, J.C. (1968). Seasonal sequence of bioluminescence and the occurrence of endogenous rhythms in oceanic waters off Woods Hole, Massachusetts. J. mar. Res. 26: 127-133

Part 2

Bioluminescence Imaging Methods

3

Bioluminescent Proteins: High Sensitive Optical Reporters for Imaging Protein-Protein Interactions and Protein Foldings in Living Animals

Ramasamy Paulmurugan
Stanford University School of Medicine, USA

1. Introduction

1.1 Bioluminescence

Bioluminescence is the production and emission of light by a living organism. Bioluminescence imaging was developed over the last decade as a tool for studying biological processes in living small laboratory animals by molecular imaging. The bioluminescence-based optical imaging is highly sensitive, low-cost, and non-invasive, enabling the real-time analysis of disease processes within the cell at a molecular level in living animals. Recent advances in protein complementation strategies have further expanded its applications by quantitatively monitoring several sub-cellular processes such as protein-protein interactions, protein dimerizations, and protein foldings. In this chapter, we provide a brief introduction to bioluminescence imaging technology and discuss its applications in studying protein-protein interactions, protein dimerizations, and protein foldings, which are some of the most important cellular processes that occur in the heart signal transduction network within the cells, by non-invasively imaging living animals.

Molecular imaging offers many unique opportunities to study biological processes in intact organisms. Bioluminescence imaging (BLI) is one of several molecular imaging strategies currently in use for studying different biological processes. It is based on the sensitive detection of visible light produced during luciferase enzyme mediated oxidation of substrate luciferin in the presence of several co-factors. The luciferase enzyme can be expressed in cells as an indicator of cellular process, and can be used to image living animals by developing tumor xenografts, or developing transgenic animals either to selectively express in a particular type of tissue using a tissue specific promoter, or in the entire animal by a constitutive promoter, to study different cellular diseases. The expressed luciferase enzyme can be imaged with an optical cooled charge coupled device (CCD) camera by injecting the substrate luciferin. Several bioluminescence reporters with a wide range of emission wavelengths are currently identified from insects and crustacean copepods **(Table 1)**. Some of the proteins were even modified by changing from a few to several amino acids by *in vitro* manipulations, and achieved considerably altered proteins with change in their emission wavelengths, which improved their detection sensitivity especially for *in vivo* imaging applications.

Bioluminescence light from firefly luciferase which emits at the ~575 nm wavelength (with several red shifted mutants) can be imaged at a depth of several centimeters within the tissues, which allows at least organ-level resolution. This technology has been applied in several studies to monitor transgene expression, progression of infection, tumor growth and metastasis, tissue acceptance/rejection in transplantation, toxicology, viral infections, and gene therapy. BLI is simple to execute, and enables monitoring throughout the course of disease, allowing localization and serial quantification of biological processes without sacrificing the experimental animal. This powerful technique can reduce the number of animals required for experimentation because multiple measurements can be made in the same animal over time, which has the added benefit of minimizing the effects of biological variation in handling different groups as control. The strengths of bioluminescence reporters are not just limited to their applications in monitoring disease progress at the cellular level. The recent development of split-reporter technology has further extended their application to monitoring sub-cellular events such as protein-protein interactions and protein-foldings that are the main focus of this chapter.

Bioluminescent Reporters	Physical Property	Source	Emission Wavelength	Substrate
Firefly Luciferase	Non-secretary	*Photinus pyralis*	575/610nm: ATP/dATP	D-Luciferin
Beetle Luciferase	Non-secretary	*Pyrearinus termitilluminans*	Red:610nm/Green: 540nm	D-Luciferin
Renilla Luciferase	Secretary	*Renilla reniformis*	482nm	Coelenterazine
Gaussia Luciferase	Non-secretary	*Gaussia princeps*	480nm	Coelenterazine
Metridia Luciferase	Secretary	*Metridia longa*	480nm	Coelenterazine
Vargula Luciferase	Secretary	*Vargula hilgendorfii*	478nm	Vargula-Luciferin
Bacterial Luciferase	Non-secretary	*Vibrio fischeri*	482nm	Fatty acids

Table 1. Bioluminescent reporters currently in use for different biological applications, and their sources and properties

1.2 Protein-protein interactions

Cells are the fundamental working units of every living system. Cells determine how a living organism functions. The complex cellular functions rely on several fundamental principles. Each cell has a nucleus that contains chemical DNA (deoxyribonucleic acid) as its genetic material, which carries all the instructions needed to direct their activities in the form of functional units called proteins. Therefore, cellular functioning ultimately depends on the performances of different proteins. Some proteins act as building blocks, such as muscle proteins, while others such as enzymes control the chemical reactions within the cells. Protein–protein interactions are important determining factors in the regulation of many cellular processes. Signaling pathways regulating cellular proliferation, differentiation, and apoptosis are commonly mediated by protein-protein interactions as well as reversible chemical modifications of proteins (e.g., phosphorylation, acetylation, methylation, and sumoylation), which normally control sub-cellular trafficking and function of proteins. To understand these modifications in proteins, and protein modification-assisted or independent protein-protein interactions, several techniques have been developed and studied in intact cells and in cell extracts. The yeast two-hybrid system is one of the earliest techniques, which used enzyme beta-galactosidase as a reporter protein at the beginning, and later was improved by adopting bioluminescent reporters for rapid measurement. The latter is used extensively in screening for protein-protein interactions and also for identifying small molecule drugs that alter (inhibit or enhance) protein-protein interactions, which can be used as therapeutic agents for treating several cellular diseases including cancer. The major limitation of this system is that it can only study the protein-protein interactions occurring in the nucleus; otherwise it requires the study proteins to be trafficked into the nucleus. The readout of yeast two-hybrid system is based on the amount of reporter proteins produced during protein-protein interaction associated transcriptional activation of reporter proteins (see more details in section 3.2). To circumvent this limitation, other techniques have been developed, including the split ubiquitin system, Sos recruitment system, dihydrofolate reductase complementation, β-galactosidase complementation, β-lactamase complementation, the G protein fusion system, and, most recently, split-luciferase (firefly luciferase, click-beetle luciferase, renilla luciferase, and Gaussia luciferase) and split-fluorescent (GFP and RFP) complementation systems. Of these, the split-luciferase complementation system provides significant advantage over other systems, particularly in measuring protein-protein interactions in cell lysates, intact cells, and cell implants in living animals by molecular imaging. The firefly luciferase complementation imaging is robust and a broadly applicable bioluminescence approach with applications in both modification-independent (phosphorylation, acetylation, methylation, and sumoylation) and dependent protein-protein interactions.

1.3 Post genomic proteomic era

We are in a post-genomic proteomic era. The completion of the human genome project has given us knowledge of the complete nucleotide sequences of human genome, their arrangements in different chromosomes, and the number of functional genes that are present in a human cell. The information collected from the human genome project along with other bio-informatic tools have led to several major new directions in science, including the characterization of RNAs (via transcriptional profiling), microRNAs, and proteins (proteomes). The human genome project estimated the number of functional genes in a human cell to range from 30,000 to 40,000. The concept of one protein, one function can accommodate only a limited number of functions, and does not explain the vastly more proteins needed by cells than those produced from the limited number of functional genes. The management of additional cellular functions, including various house-keeping functions and other specialized functions, mainly depends on the functional organ or tissue types to which these cells are part of. It is logical and even necessary to postulate that multifunctional proteins within the cell, and/or various collaborative interactions between proteins, are needed as molecular machines to carry out the work within a cell. To illustrate, the proteomes are much more dynamic and complex than the genome; it changes during development in response to external stimuli, and form large interaction networks through which they support and regulate each other. The genetic blueprint and the genome of human cells are well known. However, the functions that genome encodes and program through which the proteins are produced by the genetic blueprint are not well understood. New research is only beginning to uncover the incredibly rich diversity of protein structure, which is much more complex than that of DNA. One new direction has sought to isolate and structurally characterize all the proteins that exist in the cell (Skolnick et al., 2000; Tucker et al., 2001). Unlike DNA, proteins have a vast repertoire of structures to carry out the diversity of functions. Once the proteins are identified and characterized, a second major challenge to find out how they assemble into the molecular machines that perform the cellular functions. Identifying all of the protein-protein interactions is fundamental for understanding the cellular processes involved in virtually all biological interactions. The collection of protein-protein interactions can be visualized as a map, in which proteins are the nodes and the circuits are the interactions. A protein-protein interaction network or map would then represent a search grid on which biological circuits are constructed (Tucker et al., 2001; Wills 2001).

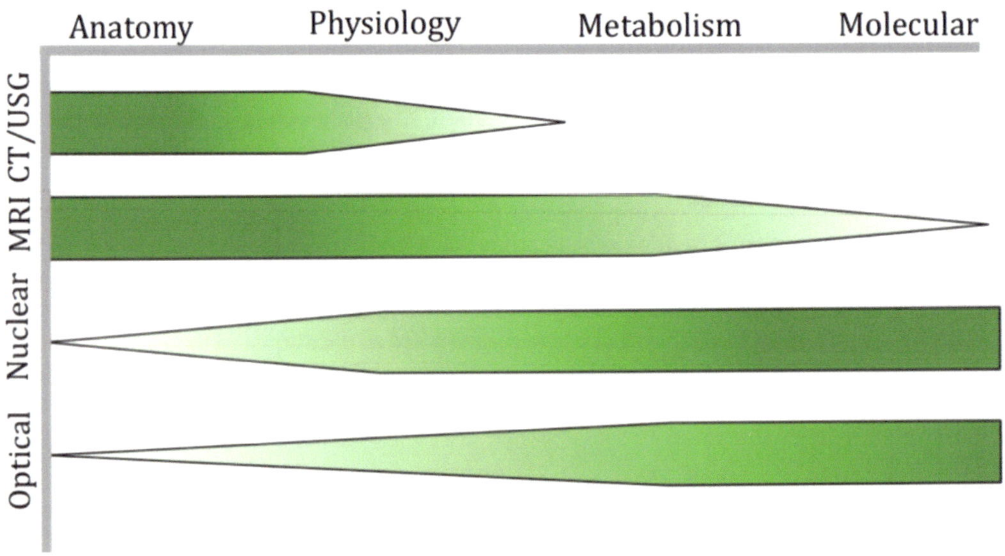

Fig. 1. Schematic illustration of current molecular imaging strategies, and their potential for providing biological informations such as anatomical details, physiological data, and metabolic status at the molecular level, for clinical applications in human. None of the current strategies is uniquely superior in independently providing different informations needed for making clinical decisions in diagnosis, staging and treatments especially in oncology, diagnosis and treatment in several other diseases; each has its strengths and weaknesses.

1.4 Complexity of protein interaction networks

There are thousands of different proteins active in a cell at any time. Many of these proteins are working as enzymes that catalyze the chemical reactions of metabolism, while others work as components of cellular machineries, such as ribosomes that read genetic information and synthesize proteins. Still proteins are involved in the regulation of gene expression. Many proteins play their functional roles only in specific cellular compartments, whereas others move from one compartment to another, acting as "signals". By directly interacting with one another, proteins continually influence other functions (Wills 2001). In addition, proteins are constantly produced and degraded in cells. The rates at which these processes occur depend on how much of each protein is already present, how they interact with each other, and with other macromolecules such as DNA and RNA, and regulate the cellular mechanisms. One protein can speed up or slow down the rate of production of another by interacting with DNA or RNA, which is needed for making that particular protein. The interactions between different proteins that control different cellular functions are therefore interdependent. When a mutation causes the loss of one of these essential protein functions, then this can significantly affect the function of many other proteins, even leading to cell death (Tucker et al., 2001). Clearly the interactions between different proteins in a cell are much more complex than previously thought, and it is vital to understand their fundamental interlinked networks. Protein–protein interactions are important determining factors in the control of many cellular processes such as transcription, translation, cell division, signal transduction, and oncogenic transformation. To modulate many of these cellular events, it is essential to delineate which proteins are involved and how they interact with one another, their precise roles in executing cellular functions, and techniques and mechanisms needed to manipulate these interactions for novel drug development or treatment strategies relevant to particular diseases. Biochemical pathways and networks require many different systems of dynamic assembly and disassembly of proteins with other proteins and nucleic acids (Michnick 2001). Much of modern biological research is concerned with how, when, and where proteins interact with other proteins involved in biological processes in the intact cellular context. The completion of the human genome project has added a major impetus in research that can provide simple approaches to study protein-protein interactions on a large scale in diseases, including cancer.

1.5 Cellular signaling pathways

The cellular regulatory mechanisms are interlinked. To understand the complex biological processes, and disease states at a molecular level, a systematic approach is necessary to illustrate signaling pathways. Efforts to elucidate the cellular mechanisms for different pathological conditions have significantly increased after the Human Genome Project. Each signaling pathway reacts to specific external stimuli that can be regulated by changes in proteins and chemicals. Recent advances in large-scale and high-throughput techniques, including functional genomics, proteomics, RNAi technology, and genomic-scale yeast two-hybrid and protein complementation assays, have provided a tremendous amount of information on signaling pathways. To extract the biological significance from the vast data, it is necessary to develop an integrated environment for a formal and structured organization of the available information, in a format suitable for analysis with bioinformatics tools. To present a signaling pathway, a database must include information on 1) the molecules involved in signaling in response to each external stimulus, 2) which direction the signal is being conveyed, and 3) how the activities and sub-cellular localizations of molecules are changed by protein modifications and/or protein-protein interactions. Analyses of the first database containing such information should made it possible to further expand the database to understand the signaling results in processes such as proliferation, differentiation, and apoptosis, and to explicate how a network can be composed of various signaling pathways in response to multiple external inputs. Signaling entities ranging from small molecules and proteins-to-protein states and protein complexes should be studied. It must be noted that these entities are not independent of one another. For instance, protein complexes are composed of proteins, and a protein binding to a small molecule can define a protein state. It is not surprising to find many gaps in the current knowledge about any particular signaling pathway. In order to organize such diverse yet incomplete information into a structured and coherent database, the use of a formal model is indispensible. Differing levels of abstraction are inter-related so that essentially the same signaling event can be described in detail at multiple levels. As model systems that implement all the parameters become available, the sharing of models with integrated biological data will be essential to fill in the gaps in our current knowledge base.

1.6 Complexity in studying protein interaction networks

There are no methods currently available to test protein-protein interaction networks, which occur within a cell without introducing a constructed system that mimics the function of its endogenous protein. It is to be expected that when a new protein of endogenous origin is introduced in a cell in addition to the level of its counterpart expressed inside a cell, it will have some direct physical effect on a number of other proteins. These new interactions may cause some changes in the functional aspects of several other proteins. Such effects can be felt right across the protein interaction network, most often becoming less significant as the distance of the new protein from the other protein increases. It is also possible for genetically modified cells to produce a new protein that will display completely new patterns of protein interactions. This may not be evident until the cells find themselves in some unusual circumstances. They may then respond in a very different way from wild-type cells. Although the genetically engineered cells may appear to behave just like wild-type cells, this cannot be guaranteed under all circumstances (Becker et al., 1990; Beeckmans 1999; Bode and Willmitzer 1975). However the techniques currently available for inserting new DNA into the chromosomes of cells do not have any specific control mechanisms, capable of directing the point of insertion in the organism's existing genome without producing significant impact on the expression level of any of the endogenous proteins. Of the gene delivery systems currently available, the adeno-associated virus is the only viral mediated vector which can normally introduce and integrate a single copy of the transgene specifically into human chromosome loci at 19 (19q13.3-qter). Otherwise, it is customary to produce millions of cells with the new DNA inserted at essentially random positions in the hope of producing at least some "hits." Screening is then conducted to find those cells, which must survive the engineering process and also express the newly inserted gene. These survivors are then subjected to further screenings to find those that seem to behave most like the wild-type, and yet possessing the new, desired, engineered properties. It is generally assumed that any harm to an organism as a result of inserting a new gene will be observed as a change in gross characteristics of the organism (Stopeck et al., 1998).

1.7 Biological importance in studying protein folding

As discussed in the previous sections, proteins are cellular macromolecules with complex structural and functional properties. Dysfunctional protein folding represents the molecular foundation of a growing list of diseases in humans and animals. Proteins undergo several levels of structural alterations executed by active chaperon complexes (e.g., Hsp90, Hsp70) and indirectly by the inherent amino acid sequences, before they become a biologically active functional entity of a cell. There is significant supporting evidence that associates the misfolding of proteins with several cellular diseases, including cancers **(Table 2)**. Biologically representative *in vitro* and *in vivo* studies of these abnormal events are best suited to the discovery of molecular mechanisms to prevent or ameliorate such diseases. There is an active search for small molecules which assist refolding of misfolded proteins into their biological functional forms, as equal or at near equal levels of native forms, for the treatment of several biochemical disorders. However, thus far no current technique can be optimally extended to imaging assays in intact living subjects. The development of novel imaging techniques to quantitatively measure the level of protein misfolding in cells and in living animals, and also of small molecule mediated refolding, will be very useful for screening and pre-clinical evaluation of drugs which rectify or cure these diseases. Normally, the conformational changes in protein folding result in the close approximation of amino and carboxy termini in a great majority of native proteins, at their functionally active forms. The 'protein folding problem' has remained one of the more perplexing quandaries in fundamental biological research ever since the classic work of Anfinsen some four decades ago on the hydrophobic-collapse mechanism. How to predict the three-dimensional, biologically active, native structure of a protein from its primary sequence, and how a protein reaches this native structure from its denatured state are still unresolved questions. The intellectual conundrum of the folding pathway of proteins, underscored by the Levinthal paradox, has been addressed to some extent over the last twenty years by various proposed mechanisms for protein folding, including the framework model (diffusion-collision and nucleation mechanisms) (Anfinsen 1973; Levinthal 1969).

There is accumulating evidence that the conditions used for refolding proteins *in vitro* are only distantly related to those found *in vivo*, where the physiological environment in living cells exerts a profound influence on protein folding owing to the involvement of the intracellular macromolecular background, which also contains folding catalysts and molecular chaperones. Aside from the relevance of the protein-folding problem to deciphering fundamental processes in cell biology, it is becoming clear that dysfunctional protein folding represents the molecular foundation of a growing list of diseases in humans and animals. There is mounting interest in such diseases arising from protein misfolding and aggregation, including Alzheimer's disease, amyloidosis, Creutzfeldt-Jakob disease, cystic fibrosis and cancer, to name a few. Molecular chaperones are involved in the protection of cells against protein damage through their ability to hold, disaggregate, and refold damaged proteins or their ability to facilitate degradation of damaged proteins. Many of the proteins implicated in the pathogenesis of misfolding diseases escape the diverse chaperoning pathways that are in place to assist and assure the fidelity of correct protein folding. More biologically representative *in vivo* structural and functional studies of these abnormal events, carried out in the context of living cell environments, are likely best suited to the discovery of molecular mechanisms to prevent or ameliorate such diseases **(Table 2)**(Goetz et al., 2003).

Disease	Mutant Protein/Protein involved	Molecular Phenotype
Inability to fold		
Cystic fibrosis	CFTR	Misfolding/altered Hsp70 and calnexin interactions
Marfan syndrome	Fibrilin	Misfolding
Amyotrophic sclerosis	Superoxide dismutase	Misfolding
Scurvy	Collagen	Misfolding
Maple syrup urine disease	α-Ketoacid dehydrogenase complex	Misassembly/Misfolding
Cancer	p53	Misfolding/altered Hsp70 interaction
Osteogenesis imperfecta	Type I procollagen pro a	Misfolding/altered BiP expression
Toxic folds		
Scraple/Creutzfeldt-jakob/familial isomnia	Prion protein	Aggregation
Alzheimer's disease	β-Amyloid	Aggregation
Familial amyloidosis	Transthyretin/lysozyme	Aggregation
Cataracts	Crystallins	Aggregation
Mislocalization owing to misfolding		
Familial hypercholesterolemia	LDL receptor	Improper trafficking
α1-Antitrypsin Deficiency	α1-Antitrypsin	Improper trafficking
Tay-Sachs disease	β-Hexosaminidase	Improper trafficking
Retinitis pigmentosa	Rhodopsin	Improper trafficking
Leprechunism	Insulin receptor	Improper trafficking

Table 2. Examples of some putative protein misfolding associated diseases and proteins involved in these diseases

2. Molecular imaging

2.1 Role of molecular imaging in cancer research

Molecular imaging, a new field of pharmacology that exploits the multidimensional approaches of light energy, has paved the way for easy understanding, interpretation, and manipulation of biological events at the molecular level. Molecular imaging is instrumental in the diagnostic aspects of various pathological conditions, and also in the evaluation of drugs which target specific molecular and biochemical processes in living cells and in intact living animals. This new field of research has flourished with the introduction of many novel molecular imaging probes, and genes which emit light either by reacting with a substrate or with induction of light waves, as well as the advancement of sensitive imaging instrumentations. Imaging techniques such as positron-emission tomography (PET), single-photon emission tomography (SPECT) with the use of radioactive tracers, and magnetic resonance imaging (MRI), are now widely used in the clinical observation of cancer pathology. Recently, the use of combinatorial techniques like PET-CT and PET-MRI are rapidly replacing conventional imaging methods. Positron Emission Tomography (PET) is an imaging technique that produces three-dimensional images of the functional processes of living subjects by capturing a pair of gamma rays emitted indirectly upon the injection of a positron-emitting radionuclide into the living body with a biomolecule.

PET was introduced by David E. Kuhl and Roy Edwards of the University of Pennsylvania in late 1950s, and has been continually updated and modified to correspond to the clinical and research needs to work as an independent or a combinatorial device (Ter-Pogossian et al., 1975). It has made invaluable contributions in cancer diagnosis, treatment, and research by revealing tumor progression both in clinical and preclinical applications, especially in the diagnosis and detection of tumor metastasis. Advances in PET scanner devices and the introduction of novel radiotracers have fueled the progress of PET imaging. Fluorodeoxyglucose (^{18}F-FDG), an analogue of glucose, is the most common radiotracer used for PET imaging, because it can reveal specific tissue metabolic activity. However, ^{18}F-FDG is phosphorylated by hexokinase and the phosphate cannot be cleared in most tissues, which can result in intense radiolabeling of tissues with high glucose uptake (Burt et al., 2001). FDG-PET is widely used in clinical oncology for diagnosis, staging and follow-up after treatment of tumors. Tissue and also molecule-specific radiotracers have been introduced to monitor the expression level of structural and functional proteins (Torigian et al., 2007). Steroid receptors have been associated with the growth of breast tumors, and thus understanding the receptor status is essential for the treatment of breast cancer. Radiolabeled ligands and their analogues are in preclinical application for receptor imaging. ^{18}F-fluro-17β-estradiol (FES) has been used in PET imaging to examine the estrogen receptor status in different tissues of living subjects (Mintun et al., 1988). Bombesin, a peptide isolated from the frog *Bombinas bombina*, binds with gastrin-releasing peptide (GRP) receptor and has been implicated in breast cancer. This property has led the development of radiolabeled bombesin for peptide receptor imaging in breast cancer diagnosis (Scopinaro et al., 2002).

Single Photon Emission Computed Tomography (SPECT) is another imaging technique that is similar to PET imaging. Unlike PET, however, the tracer used in SPECT emits gamma rays that can be measured directly. In SPECT, the 2-D view of 3-dimensional images is acquired by a gamma camera and eventually 3-D data set is generated with the use of computer based tomographic reconstruction algorithm. Magnetic Resonance Imaging (MRI) is another widely used imaging technique, but unlike PET and SPECT, MRI can be used to view the anatomical nature of living subjects by generating data about the functional status of tissues (MacDonald et al., 2010). Magnetic Resonance Imaging (MRI) is a well-established diagnostic method to detect cancer. It has been widely used to detect breast cancer as it produces the highest sensitivity in spatial resolution of all imaging modalities. Guinea et al. (2010) investigated and analyzed the possible relationship between the magnetic resonance imaging (MRI) features of breast cancer and its clinicopathological and biological factors such as estrogen and progesterone receptor status, and expression of p53, HER2, ki67,

VEGFR-1, and VEGFR-2 (Fernandez-Guinea et al., 2010). Their results did not show a significant association between the MRI parameters and any of the biological factors included in the study. By contrast, other reports have shown that high spatial dynamic MRI of morphological or kinetic analysis are associated with prognostic factors such as expression of estrogen receptor (ER), expression of progesterone receptor (PR), and expression of p53, c-erbB-2, and Ki-67 (Fernandez-Guinea et al., 2010, Szabo et al., 2003) found that rim enhancement pattern, early maximal enhancement, and washout phenomena are associated with the poor prognostic factors of histological differentiation, high Ki-67 index, and negative PR expression status (Szabo et al., 2003). It was suggested that these MR images could be useful in the prognosis of breast cancer. The MR signals may also be used to noninvasively identify highly aggressive breast carcinomas and help differentiate between benign and malignant lesions. The difference in contrast enhancement has been associated mainly with a higher vascular permeability in tumors, and the overexpression of c-erbB2 in tumor cells is closely linked to increased expression of vascular endothelial growth factor (VEGF) and Ki-67 in proliferation (Szabo et al., 2003).

Similar to nuclear and magnetic imaging modalities, optical imaging has also made a sizable contribution in medical imaging (Weissleder and Pittet 2008). Fluorescence and bioluminescence methods are used as a source of contrast in optical imaging, but a major setback in optical imaging is the lack of penetration depth, which prevents its wide clinical applications in humans. Near infrared (NIR) imaging has been identified as a useful optical imaging technique because of the lower absorption coefficient of tissue to light in near infrared region. Radiolabeled antibodies have been evaluated for breast cancer diagnosis since 1978 with tumor associated antigens such as carcinoembryonic antigen (CEA) and the polymorphic epithelial mucin antigen (MUC1) (Goldenberg et al., 1974). They were widely used in studies of immunolocalization and radioscintigraphy. Clinically, affinity purified ^{131}I-labeled goat anti-CEA IgG was subjected in selective breast tumor targeting (Goldenberg and Sharkey 2007). CEA-Scan with Arcitumomab, an US FDA approved anti-CEA antibody, could detect breast cancer that has been missed by mammography. In addition, antibodies against the HER2/neu receptor have also been investigated in detail either as therapeutic and/or diagnostic agent. Biotinylated anti-HER2/neu antibodies have been used to increase the contrast of MR images when they bind with avidin-gadolinium complexes (Artemov et al., 2003). The ^{111}In labeled Trastuzumab (Herceptin) Fab has been identified as a selective imaging agent to localize HER2/neu receptor in small BT-474 tumors (Tang et al., 2005).

2.2 Reporter gene imaging in living animals

Molecular imaging is a rapidly expanding field that attempts to visualize fundamental molecular/cellular processes in living subjects (Gambhir 2002; Massoud and Gambhir 2003; Weissleder 2002). Imaging molecular events in cells in their native environment within the living subjects probably result in the least amount of perturbation of normal signaling processes. However, this advantage of non-invasive imaging has a trade-off. For example, most current techniques do not have single cell resolution at any significant depth within the animal. Instead, bulk signals from large numbers of cells (hundreds to millions) are needed. Newer methods that allow the observation of single cells within living subjects are under active investigation, but are more invasive in nature (Jung and Schnitzer 2003; Mehta et al., 2004). To produce a signal detectable outside the animal subject, the cells located inside the subject must produce a signal of sufficient intensity. The signal may come from a fluorescent protein excited at the correct wavelength, the interaction of a bioluminescent protein with its substrate **(Figure 2)**, or from radiolabeled substrates that emit a signal in the form of gamma rays. For optical signals, red light and near infrared light have the best tissue penetration, and are therefore preferred. For radiation-based signals, the use of single photon emitters and positron emitters generating gamma rays are favored. It is not possible to use beta emitters (e.g., ^3H), due to their minimal tissue penetration. The focus of this chapter is primarily on optical technologies for imaging protein-protein interactions. Although other approaches such as microPET can be used for imaging protein-protein interactions (Luker et al., 2002a; Massoud et al., 2010), the much lower cost, higher throughput, and greater sensitivity of optical imaging in small animals favor its use for imaging protein-protein interactions. Additional discussions of other small animal imaging technologies including microPET may be found elsewhere (Cherry and Gambhir 2001; Massoud and Gambhir 2003).

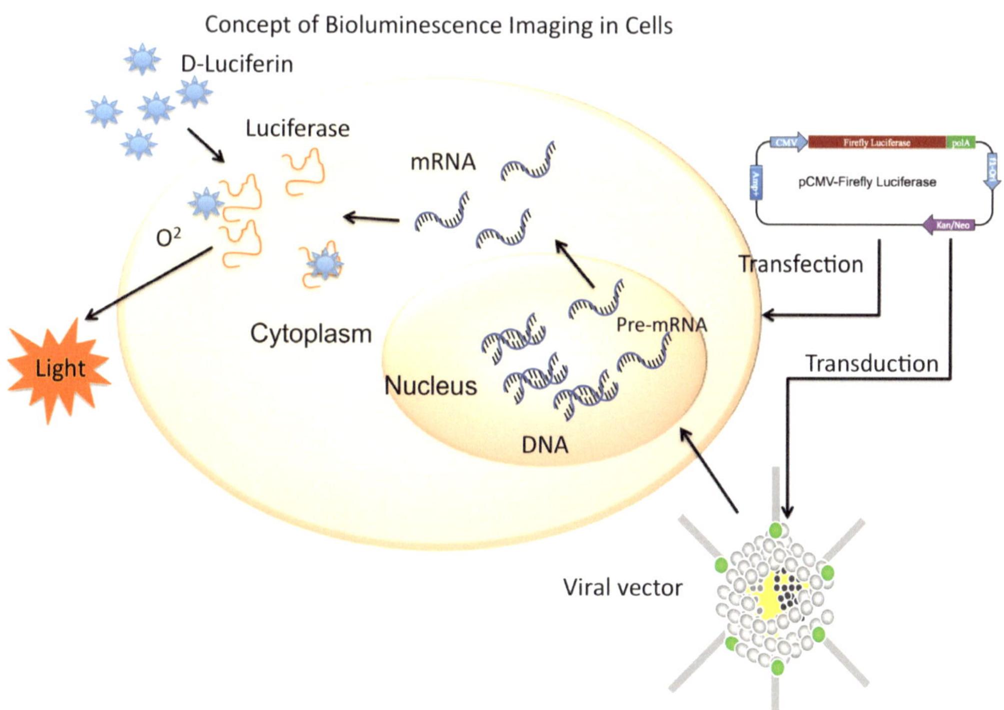

Fig. 2. Schematic illustration of the principle of optical bioluminescence imaging in cells. In this strategy mammalian cells are labeled to express bioluminescent protein under a constitutive (CMV, LTR, Ubiquitin, or CAG) or tissue-specific or an inducible promoter, either by transfecting a plasmid with chemical agent (Liposome) or transduced with a viral vector (Lentivirus, Adenovirus, Retrovirus, or Adeno-associated virus). The cells can be allowed to express luciferase protein for a particular period of time and imaged by exposure to substrate luciferin in intact cells, or can be measured by luminometer in cell lysates by adding luciferin and other co-factors.

There are two primary types of optical imaging systems for living subjects: a) fluorescence imaging, which use emitters such as green fluorescent protein (GFP), wavelength-shifted GFP mutants, red fluorescent protein (RFP), "smart" near-infrared fluorescent (NIRF) probes, and b) bioluminescence imaging, which utilizes a specific enzyme-substrate reaction such as Firefly luciferase/D-Luciferin, Renilla luciferase/coelenterazine (Bhaumik and Gambhir 2002a; Contag and Ross 2002; Tung et al., 1999) and several other bioluminescent proteins with the respective substrates (Substrates and properties are shown in **Table 1)**. Emission of light from fluorescent markers requires external light excitation, while bioluminescence systems generate light *de novo* after an injectable substrate is introduced. In both cases, emitted light can be detected with a thermoelectrically cooled charge-couple device camera (CCD), which can detect light in the visible light range (400 nm to 750 nm) to near-infrared range (~800 nm) **(Figure 3).** Cooled to -120 to -150°C, these cameras are exquisitely sensitive to even weak luminescent sources within a light-tight "black-box"

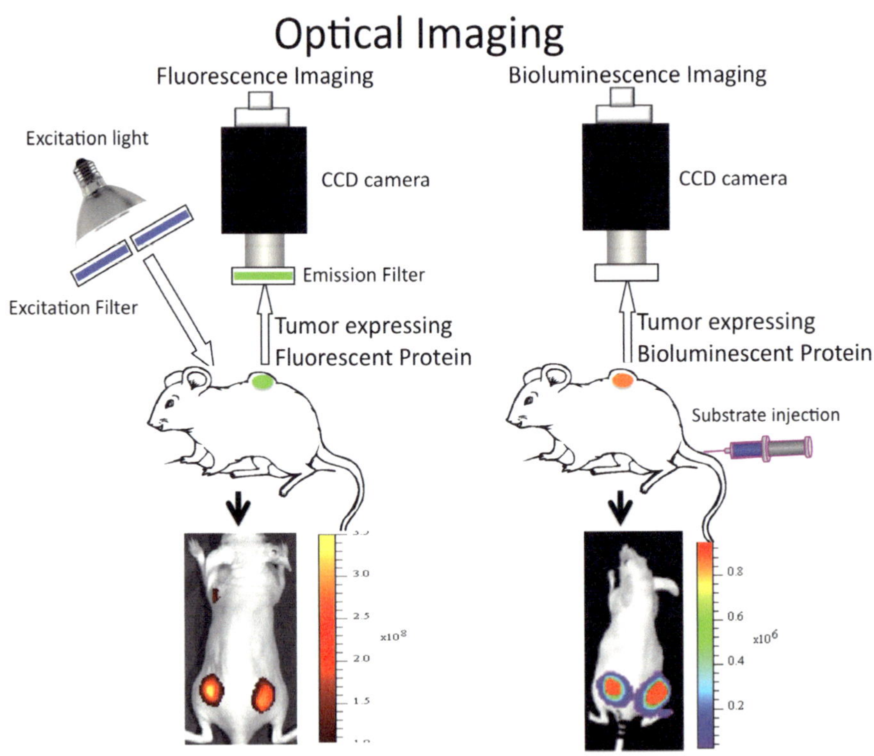

Fig. 3. Scheme of optical imaging (bioluminescence and fluorescence) in living animals. In this strategy, mammalian cells stably expressing bioluminescent or fluorescent proteins are implanted in animals (orthotopic or xenograft) and allowed to grow the tumors. The animals were imaged by exciting with respective excitation wavelength of the protein used for labeling in fluorescence imaging. The emitted light was captured by an optical cooled charge coupled device camera, and quantitated by compatible software provided with the system. Similarly, in bioluminescence imaging, the animals were injected with the respective substrate of the bioluminescent reporter used for labeling the cells, and light emitted from the tumor cells is collected without passing through filters.

chamber, allowing for quantitative analysis of the data. The method of imaging bioluminescence sources in living subjects with a CCD camera is relatively straightforward: the animal is anesthetized, injected intravenously or intraperitoneally with the substrate, and placed in the light-tight chamber for a few seconds to minutes. A standard light photographic image of the animal is obtained, prior to a bioluminescence image captured by the cooled CCD camera positioned above the subject within the confines of the dark chamber. A computer subsequently superimposes the two images on one another, and relative location of luciferase activity is inferred from the composite image. An adjacent pseudocolor scale indicates relative or absolute number of photons detected. This scale does not reflect the color (wavelength) of the emitted photons, but only the number of such photons, measured in relative light units per minute (RLU/min). At present, the primary disadvantage of fluorescence as compared to bioluminescence is the background level of auto-fluorescence of tissues in the former approach. However, methods are being developed to correct for auto-fluorescence and may allow greater use of fluorescence in the study of protein-protein interactions in living subjects. To date, the initial validation of *in vivo* imaging has been based primarily on bioluminescence. If auto-fluorescence issues can be minimized, then it may be possible to image multiple interactions using different fluorescent reporters. Comparison of optical-based imaging systems with the other imaging modalities, such as the radionuclide-based or MRI-based systems, reveals important differences. One major advantages of optical-based reporter systems is that they are at least an order of magnitude more sensitive than the radionuclide-based techniques at limited depths (Ray et al., 2003). Furthermore, the direct and indirect costs of optical systems are generally less than radionuclide-based techniques or MRI. However, significantly less spatial information is obtained from optical imaging, and the signal obtained from light-emitting reporter systems is limited by the tissue depth from which it arises. Furthermore, while significant progress has been made to localize fluorescent signals tomographically to obtain distribution of fluorochromes in deep tissues (Ntziachristos et al., 2002), there are currently only prototype instruments to obtain three-dimensional localization of the targeted optical probes.

2.3 Optical reporter genes to image cellular process in small animal models

Luciferases are enzymes that emit light when they react with a specific substrate. A diverse group of organisms make use of luciferase-mediated bioluminescence. Luciferases that catalyze the light emitting reactions of fireflies, coelenterates, or bacteria show no nucleotide homology to each other. The substrates (i.e., luciferins) of these reactions are also chemically unrelated (Wilson and Hastings 1998). All these bioluminescent reactions combine molecular oxygen with luciferin, to form a luciferase-bound peroxy-luciferin intermediate. This, in turn, releases photons of visible light (Wilson and Hastings 1998) over an emission spectrum range between 400 and 620 nm. Experimentally, the emitted light is used as a "reporter" for the activity of any regulatory elements that control expression of luciferase. Firefly luciferase (FLUC), cloned in 1985 from the firefly *Photinus pyralis*, is now emerging as the gene of choice for *in vivo* and *in vitro* reporting of transcriptional activity in eukaryotic cells (de Wet et al., 1985). FLUC emits light from green to yellow in the presence of D-Luciferin, ATP, magnesium, and oxygen. The short half-life and fast rate of turnover of FLUC ($T_{1/2} \sim 3$ h) in the presence of D-Luciferin allows for real-time measurements, because the enzyme does not accumulate intracellularly to the extent of other reporters; thus, the relationship between the enzyme concentration and the intensity of emitted light *in vitro* is linear up to 7-8 orders of magnitude. These properties potentially allow for sensitive non-invasive imaging of FLUC reporter gene expression in living subjects (Massoud et al., 2004; Wu et al., 2002). In recent years, considerable work with non-invasive imaging of firefly luciferase has been carried out (Bhaumik and Gambhir 2002b; Contag et al., 2000; Contag et al., 1997; Contag et al., 1998; Jawhara and Mordon 2004; Paulmurugan et al., 2002b; Ray et al., 2002a; Wu et al., 2002; Wu et al., 2001).

The use of a second bioluminescent reporter [(Renilla Luciferase (RLUC)] with different substrate utility than firefly luciferase, has allowed for monitoring of more than one process at a time in mammalian cells. Renilla luciferase (RLUC), originally cloned and sequenced from the sea pansy, *Renilla reniformis*, by Lorenz et al. (Lorenz et al., 1991), has been used as a marker of gene expression in bacteria, yeast, plant and mammalian cells (Lorenz et al., 1996). RLUC is widely distributed among coelenterates, fishes, squids and shrimps (Hastings 1996). The enzyme RLUC catalyzes oxidation of its substrate, coelenterazine, leading to bioluminescence. Coelenterazine has an imidazolopyrazine structure [2-(p-hydroxybenzyl)-6-(p-hydroxyphenyl)-8-benzylimidazo [1, 2-a] pyrazin-3- (7H)-on], and upon oxidation, releases blue light across a broad range, peaking at 480 nm (Wilson and Hastings 1998). However, the native RLUC protein has some inherent limitations when used in mammalian cells. Ten percent of its codons are associated with poor translation in mammalian cells, limiting expression efficiency. Also, the presence of a large number of potential transcription factor binding sites within RLUC sequences can cause anomalous transcriptional behavior in mammalian cells. For many *in vivo* imaging applications, researchers have utilized a synthetic *Renilla* luciferase reporter gene (hRLUC) that has been codon optimized for efficient translation in mammalian cells. In addition, deletion of poly (A) signals (AATAAA) and incorporation of a Kozak sequence at the beginning of the gene contributed to a better expression. The resulting reporter gene has a higher transcriptional efficiency, which enhances the detection of the reporter enzyme in cell culture and living animals (Bhaumik et al., 2004).

After a long lag time, recently several other luciferases with different properties have been isolated. Gaussia luciferase (GLUC) with its secretary property has been recently used for secretary biomarker identification and validation in human cancers. As GLUC uses the same substrate (coelenterazine) as RLUC and emits light in a similar wavelength (480nm), but differs in its secretary property, its use has been limited in several applications. Similarly, Metridia luciferase, with similar substrate utilization, emission wavelength and secretary property as GLUC, has been identified, but it has not added much to the field of bioluminescence imaging in living animals. Recent isolation of another luciferase (Vargula luciferase) with similar physical properties (secretary, emits light at 480nm) as GLUC, but uses a different substrate (Vargula luciferin), has the advantage of adding another bioluminescent protein for multiplexing different bioluminescent reporters to enable simultaneous monitoring of several biological processes in cells and in living animals.

3. Bioluminescent assays to study protein-protein interactions

As we discussed briefly in section 1.6 the different assays currently available for studying protein-protein interactions, in this section we further explain how these assay systems work, and discuss the advantages most of these assays have over other complementary systems, and their contributions to the field of molecular imaging in disease monitoring and drug development processes.

3.1 Yeast two-hybrid system to study protein-protein interactions

Protein-protein interactions play vital part in almost all biological processes. Large networks of interacting proteins control many important regulatory pathways. A full understanding of any pathway or cellular processes will require a map of the binary interactions among the proteins involved. One of the most widely used methods to detect biologically important protein-protein interactions is the yeast two-hybrid system (Fields and Song 1989). In a two-hybrid assay, the two proteins are expressed in yeast one fused to a DNA-binding domain (BD) and the other fused to a transcription activation domain (AD). If the two proteins interact, they activate transcription of one or more reporter genes that contain binding sites of the BD. Investigated first in yeast, this classical two-hybrid system was later adopted for mammalian cells with certain modifications (Luo et al., 1997). When expressed simultaneously in the same cell, the interactions between the two mammalian proteins bring the activation domain (VP16) and binding domain (Gal4) together that in turn bind to the Gal4 domain binding sequences followed by a reporter gene. High levels of reporter gene expression will indirectly indicate the physical interactions between the proteins of interest. Most of the approaches used for high-throughput two-hybrid studies have been limited to *in vitro* assays and cultured cells that do not represent the actual scenario in intact animal, as well as being unable to predict the kinetics of the interaction. Thus, the development of methods to non-invasively and repetitively image protein-protein interactions using the mammalian two-hybrid system in living animal would significantly increase our knowledge on the intricacy and complexity of different regulatory pathways.

To overcome these problems, we and others have developed a modified mammalian two-hybrid system to detect protein-protein interactions in living mice using bioluminescence and positron emission tomography (PET) imaging techniques (Luker et al., 2002b; Luker et al., 2003; Ray et al., 2002b). To demonstrate the use of this system for imaging in living animals, the two known interacting proteins, ID and MyoD, were used. These two proteins strongly interact *in vivo* during muscle generation (Ray et al., 2002b). To modulate the expression of these two fusion proteins (ID-GAL4 and MyoD-VP16), we used the NF-kB promoter to drive expression of the Id-gal4 and/or myoD-vp16 fusion genes, and utilize TNF-α to induce the NF-kB promoter that controls the expression level of fusion proteins in response to TNF-α dose. The reporter construct comprised of five GAL4 binding sequences followed by the firefly luciferase reporter gene. In cell culture, co-transfection with the effector and reporter plasmids with variable levels of expression regulated by TNF-α show induction in FLUC activity. The FLUC activity directly correlate with the interaction between the proteins used, which are Id and myoD in this case. By replacing these two proteins, it is possible to look for other proteins that are currently not known for their interactions **(Figure 4)**.

A similar system was developed for a tetracycline inducible bi-directional vector carrying two other proteins (the tumor suppressor p53 gene and SV40-T antigen) by other groups (Luker et al., 2002b; Luker et al., 2003). Interactions of p53 and SV40-Tag proteins after doxycycline induced expression resulted in formation of the VP16-Gal4 trans-activator complex that binds to the Gal4 binding sequences driving expression of a HSV1-sr39TK-GFP reporter fusion protein. Expression of GFP (fluorescence imaging) is detected at the level of single cell, and expression of HSV1-sr39TK is imaged in living mice with microPET. Clones of HeLa cells stably expressing both the reporter plasmid and the bi-directional effector plasmid are isolated, and levels of interacting proteins are measured with increasing doses of doxycycline and further confirmed by western blots, thymidine kinase radiotracer assays, and fluorescence microscopy. Promising cell culture studies have led to further investigations into the protein-protein interactions in living mice using different modalities in cells and *in vivo* in living animals.

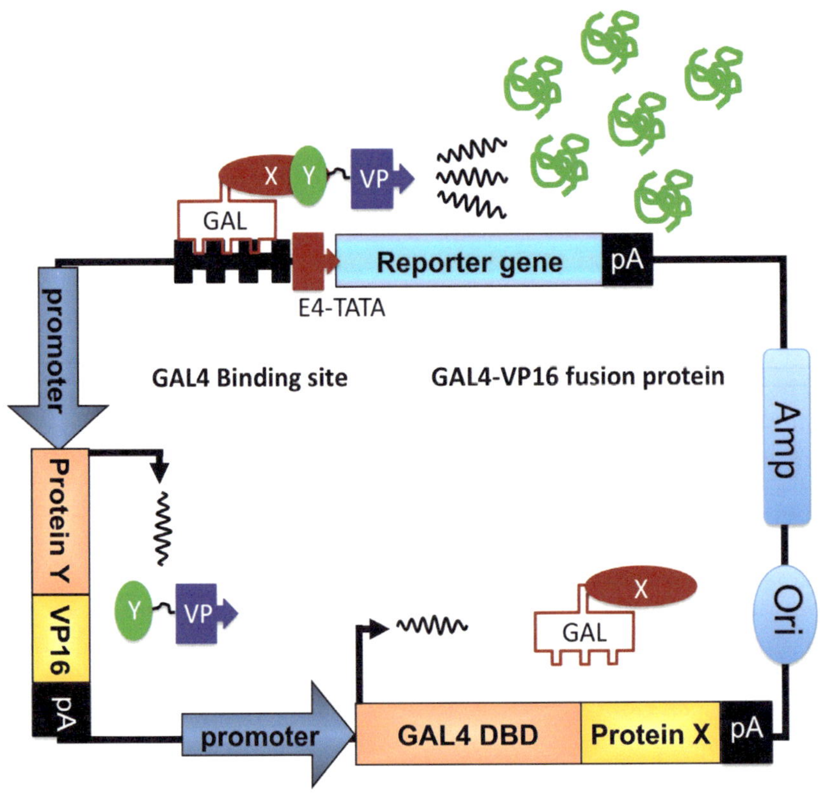

Fig. 4. Scheme of a single vector mammalian two hybrid system to study protein-protein interactions. The vector is designed to have several components needed for measuring protein-protein interaction. In this strategy, the vector expresses two fusion proteins under two separate promoters. They can be either constitutive or inducible. One fusion protein contain Gal4-DNA binding domain (Gal4-DBD) expresses with one of the study protein (Gal4-DBD-Protein-X), and the other one expresses a small transactivating peptide derived from herpes simplex virus thymidine kinase (HSV1) with another protein (Protein-Y-VP16). The same vector has a specific DNA sequence repeated five times on which the Gal4-DBD can bind, and a minimal promoter (TATA box), followed by a reporter gene (Luciferase). This minimal promoter will not express any reporter protein until the Gal4-DBD binds to it and brings the VP16 domain fused to another fusion protein by protein-protein interactions. The amount of luciferase expression directly relate to the interaction which occurs between proteins X and Y.

However, both the two-hybrid (TNF-α and tetracycline inducible) approaches have several limitations. Both use strong and constitutively (spontaneous) interacting proteins and are unable to fully address weakly associated proteins or proteins with differential binding affinity, and also to decipher the time-kinetics of protein-protein interactions. Moreover the two-hybrid system could only detect the interacting proteins in the nucleus, not in the cytoplasm where the largest pool of protein-protein interactions responsible for many regulatory pathways occur. It is hoped that by combining non-invasive imaging approaches involving detection of cytoplasmic protein-protein interaction (through protein complementation study, split reporter strategy, FRET/BRET study etc.) with the two-hybrid system, it will be possible to measure the complete spectrum of the pharmacokinetics and pharmacodynamics of protein interactions in different regulatory pathways, as well as to perform screening and pre-clinical evaluation of small molecule drugs for therapeutic applications.

3.2 Bioluminescent reporter protein complementation assays to study protein-protein interaction

Protein complementation assays with split luciferases (split Firefly, split Renilla, and split-Gaussia luciferases) are highly useful techniques for studying protein-protein interactions. Functional proteins can be assembled from one or more non-covalently attached polypeptides, with the efficiency of assembly a measure of real time protein-protein interactions, both in cells and in living animals. As discussed before, β-galactosidase from *Escherichia coli* is one of the first enzymes used extensively as an experimental reporter, long before the discovery of several other reporter proteins such as Chloramphenicol Acetyl Transferase (CAT), luciferases, and fluorescent proteins. Active β-galactosidase is a tetramer that hydrolyzes terminally non-reduced β-galactose residues in sugars, glycoproteins, and glycolipids (Hucho and Wallenfels 1972; Johnsson and Varshavsky 1994a; Johnsson and Varshavsky 1994b; Loontiens et al., 1970; Nichtl et al., 1998; Stagljar et al., 1998). The identification of the α-complementation process of β-galactosidase opened up the idea of using protein fragments coupled with an enzymatic assay to gauge protein interactions (Hodges et al., 1992; Ullmann et al., 1968). Each monomer of the tetramer can be cleaved into a small N- terminal α-fragment (50-90 residues) and a large (135 kDa fragment) ω-fragment. Addition of purified α-fragment to dimers of enzymatically inactive purified ω-fragments forms an active tetrameric enzyme, in a process called α-complementation (Hodges et al., 1992; Kippen and Fersht 1995; Smith and Matthews 2001; Ullmann et al., 1968). This process suggests that synthetically separated fragments of a single polypeptide might complement each other, and give rise to an enzymatically active protein, particularly if the interaction is aided by fusion of the halves to strongly interacting moieties. In the "split protein" strategy, a single reporter protein/enzyme is cleaved into N-terminal and C-terminal halves; each half is fused to one of two interacting proteins, X- and Y. Physical interactions between the two proteins reconstitute the functional reporter protein, leading to enzymatic activities that can be measured by *in vitro* or *in vivo* assays. This split protein strategy can work either through protein-fragment complementation assays, or intein-mediated reconstitution assays (Ozawa et al., 2001). To date, a number of different reporter proteins (β-lactamase, β-galactosidase, ubiquitin, dihydrofolate reductase, firefly luciferase, renilla luciferase, Gaussia luciferase, etc.) have been adapted for split-protein strategies. Advantages of this approach include its ability to monitor the interaction of proteins occurring in their specific cellular compartments, utility for drug evaluation, low background, ability to measure real-time interactions, applications in protein array analysis, and potential adaptability to study both cells and in living animals. However, not every reporter protein can be used for this strategy (Ozawa et al., 2001; Paulmurugan et al., 2004; Pelletier et al., 1999; Remy and Michnick 1999; Rossi et al., 1997; Wehrman et al., 2002) **(Figure 5)**.

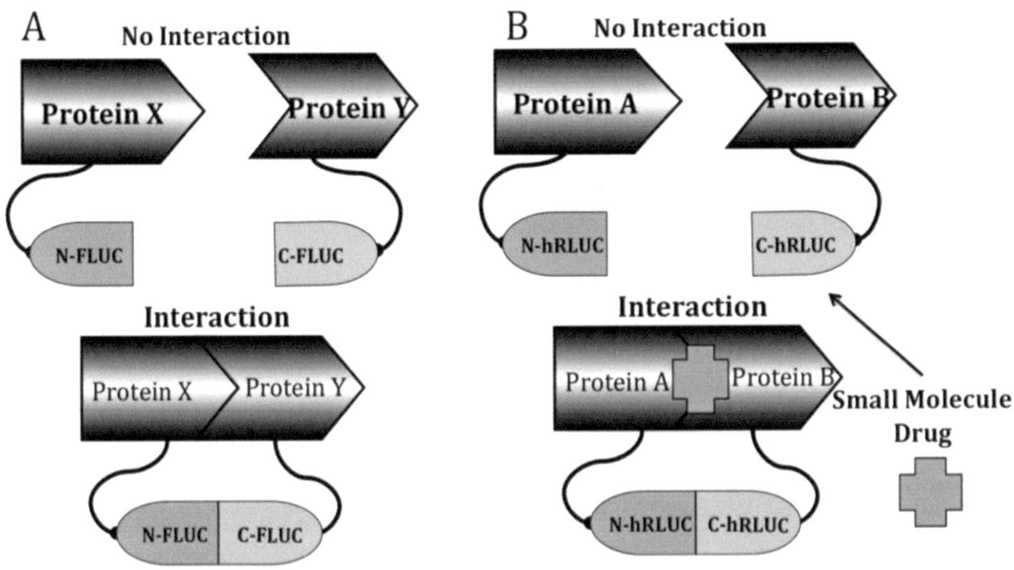

Fig. 5. Scheme of a split-reporter protein complementation system for studying protein-protein interactions. **(A).** ***Complementation system to study spontaneous protein-protein interactions***. In this strategy, N- and C- terminal firefly luciferase reporter protein fragments are attached to two study proteins and are expressed as individual fusion proteins (N-FLUC-Protein-X and Protein-Y-C-FLUC). When cells are transfected to co-express these two fusion proteins, the interaction between protein-X and protein-Y brings the N-FLUC and C-FLUC fragments to a close proximity and induces the complementation to produce reporter signal. **(B).** ***Complementation system to study small molecule mediated protein-protein interactions.*** In this strategy N- and C- terminal Renilla luciferase reporter protein fragments are attached to two study proteins and are expressed as individual fusion proteins (N-hRLUC-Protein-A and Protein-B-C-hRLUC). When cells are transfected to co-express these two fusion proteins, the interaction between protein-A and protein-B is induced by a small molecule drug, which brings the N-hRLUC and C-hRLUC fragments to a close proximity and causes the complementation to produce reporter signal. In both strategies, the amount of luciferase signals produced through complementation directly relates to the interaction which occur between proteins X and Y or A and B (Paulmurugan and Gambhir 2003; Paulmurugan et al., 2009).

We made several combinations of N- and C- terminal fragments of the FLUC, RLUC, and GLUC enzymes by a semi-rational dividing approach and used these fragments for protein-protein interactions and protein folding studies. These fragments have been efficiently used in complementation assays for the detection of insulin-mediated phosphorylation, as gauged by the subsequent interaction of insulin receptor substrate peptide and its interacting partner SH2 domain of PI-3kinase (Ozawa et al., 2001). They have also been used to detect the interactions of the myogenic differentiation proteins Id and MyoD, in both cell culture and non-invasive repetitive optical imaging in living mice (Paulmurugan et al., 2002b). Separately, Luker et al., have also described a systematic truncation library yielding alternative complementary N- and C- fragments of FLUC. These fragments were used to monitor rapamycin-mediated interactions of rapamycin binding proteins (Luker et al., 2004). We also have used our FLUC fragments to study Rapamycin-mediated interactions, and found the complementation to be too weak for imaging in living animals by optical CCD camera.

RLUC and GLUC are the smallest optical bioluminescent reporter proteins, and we have identified several split sites for these proteins and found selective combinations that are efficient for studying protein-protein interactions through a protein fragment complementation strategy. This reporter protein, when rationally split (RLUC: between residues 229 and 230 or 235 and 236; GLUC: between 105 and 106) (Paulmurugan and Gambhir 2003), functions efficiently in both cell culture and in living animals, as we have demonstrated with several different protein partners. Fragments generated by splitting between residues 229 and 230 for RLUC and 105 and 106 for GLUC, were used to study rapamycin-induced interactions of human proteins FRB (FKBP12 Rapamycin Binding domain) and FKBP12 (FK506 Binding protein) (Paulmurugan et al., 2004), and also the inherent homodimerization property of mutant HSV1-sr39TK (Massoud et al., 2004). One limitation associated with the use of both RLUC and GLUC is their relatively rapid reaction kinetics, requiring early time-point measurements (Bhaumik and Gambhir 2002a). Nevertheless, this split reporter system appears highly suitable for studying protein-protein interactions in cell culture and in living animals owing to its strong optical bioluminescence, generating a signal that is amplifiable through an enzymatic process.

3.3 Bioluminescent reporter protein complementation assay to study protein folding

As we discussed in section 1.7, techniques for studying protein foldings are crucial in measuring the functionality of important biologically active proteins. Several experimental techniques are currently available for *in vivo* evaluation of protein folding within intact cells, mainly relying on intramolecular fluorescence resonance energy transfer (FRET) (Morris et al., 1982; Russwurm et al., 2007; Tsien 2009) of labeled residues or attached variants of green fluorescent protein (GFP), but none of these techniques can be optimally extended to imaging assays in intact living subjects. Protein complementation assays based on bioluminescent reporters are highly sensitive in measuring not only protein-protein interactions, they can also be efficiently used for studying protein foldings in intact cells and in living animals by imaging. The application of protein complementation assays based on bioluminescent reporters are more generalizable for measuring protein folding, but can be used for imaging of protein folding in intact living subjects. The ability to detect, locate, and quantify protein folding in the setting of a whole living animal model has important implications: (1) to characterize the functional aspects of the fundamental process of protein folding, including the study of biologically relevant and important factors, within realistic and relatively undisturbed confines of cells that are also present in the midst of fully functional and intact whole-body physiological environments; and (2) to accelerate the evaluation in living animal models of emerging novel classes of drugs that promote folding and conformational stability of proteins (e.g., those directed at the molecular chaperone heat shock protein 90 (Hsp90)). Emerging strategies for molecular imaging of biological processes in living small animal models of disease offer many distinct advantages over conventional *in vitro* and cell culture experimentation.

In conclusion, we previously identified suitable split sites in the molecule of hRLUC that generated an N- terminal 229-residue fragment (N-hRLUC) with minimal independent activity and an inactive C- terminal 82-residue fragment (C-hRLUC) of the reporter protein. Together, they were able to produce significant recovered activity through assisted complementation. Later we used these reporter fragments along with reporter fragments of firefly luciferase (FLUC) to test protein-protein interactions mediated in different experimental settings: 1) proteins that spontaneously interact; 2) proteins that interact when a small molecule is present; 3) protein-protein interactions that are blocked by small molecules; 4) proteins-protein interactions that are mediated by phosphorylation; and 5) protein hydroxylation that is mediated protein-protein interactions **(Table 3)**. After a thorough investigation of the system for studying protein-protein interactions, we demonstrated that the normal conformational changes during protein folding, which result in the close approximation of amino and carboxy termini of the proteins, can be measured by using intramolecular complementation of correctly oriented chimeric split imaging reporters in a strategy to detect, locate, time, quantify, and image protein folding in living subjects. As we have extensively studied ligand induced estrogen receptor (ER) folding in cells and in living animals, the following section will explain in detail the intramolecular folding of ER in response to different steroidal and non-steroidal ligands in cells and in living animals.

3.4 Protein complementation assays to study ligand- induced estrogen receptor folding

Estrogens are responsible for the growth, development, and maintenance of the reproductive, skeletal, neuronal, and immune systems as well as several other systems of the body. The physiological effects of these hormones are mediated by the estrogen receptor (ER), which is a ligand-inducible nuclear transcription factor (Tsai and O'Malley 1994). In the classical pathway of steroid hormone action, 17β-estradiol (E2), hormones, and a variety of other estrogens bind to the ligand-binding domain (LBD) of ER, leading to its dimerization and subsequent binding to a specific regulatory sequence in the promoters of ER target genes known as the estrogen response elements (Gronemeyer 1991; Schuur et al., 2001) that then trigger activation or repression of many downstream target genes (Brzozowski et al., 1997). The deficiency or excess of estrogens can lead to various pathological conditions, including osteoporosis and breast carcinomas (Beck et al., 2005), making ER a major cellular therapeutic target. ER activity in regulating target genes is modulated by the binding of both steroidal and synthetic non-steroidal ligands. The ligand binding with ERs induces various conformations that control their interactions with transcriptional co-regulators. Estrogen receptors (ERα and ERβ) regulate the expression of a number of gene products required for the growth of cells in response to the endogenous estrogen 17β-estradiol (E_2). The crystal structures of ER ligand-binding domains (ER-LBDs) complexed with different ER ligands provide useful insight into the design and synthesis of new ligands. Although computer modeling and structure-based design can help predict molecular interactions and structure-activity relationships, the pharmacological actions of these ligands are unpredictable and require further biological evaluation. Thus, it remains important to fully characterize nuclear hormone receptor ligands in cells and in animal models before considering their use in humans. Several assay systems are currently available to characterize ER ligands for their biological activity through ER *in vitro* and in cell-based assays, but only a few can be directly extended for use in animals. We approached this issue by using bioluminescence imaging to study estrogen biology in living animals. Our system involves monitoring luminescence that derives from intramolecular complementation of a split luciferase gene that is activated by ligand-induced folding of the ER-LBD, and the responsiveness of various versions of this ligand sensor system to selected ER ligands was validated in cellular systems.

Split-Reporter Proteins	Imaging Modality	Protein-Protein Interactions	Nature of interaction
Firefly Luciferase (FLUC)	Bioluminescence	Id/myoD	Spontaneous interaction
Renilla Luciferase (RLUC)	Bioluminescence	FRB/FKBP12	Rapamycin mediated interaction
Gaussia Luciferase (GLUC)	Bioluminescence	pVHL/HIF1-α	Hydroxylation mediated interaction
Green Fluorescent Protein (GFP)	Fluorescence	Ras/Raf	Phosphorylation mediated interaction
Red Fluorescent Protein (RFP)	Fluorescence	Hsp90/p23	Spontaneous interaction
Thymidine Kinase (HSV1-TK)	PET	p53/SV40-Tag	Spontaneous interaction
		ER/ER	Estrogen mediated homodimerization
		TK/TK	Spontaneous dimerization
		ER-Folding	Estrogen mediated intramolecular folding
		IRS1/SH2	Phosphorylation mediated interaction
		c-myc/GSK3β	Phosphorylation mediated interaction

Table 3. List of optical and PET based split reporter complementation systems developed and protein-protein interactions and protein foldings studied

The crystallographic studies with ER–LBD have shown that conformation of helix 12 (H12) is critical in responses observed with various ER ligands (Brzozowski et al., 1997; Pike et al., 2000; Shiau et al., 1998). The conformation of H12 behaves as a "molecular switch" that either prevents or enhances the binding of ER to an array of co-activator proteins, which then activates transcription of many downstream estrogen-regulated genes responsible for cell growth. Given the critical role of H12 in ER signaling, we reasoned that it might be feasible to develop an intramolecular ER folding sensor with specific split reporter complementation patterns to study ligand pharmacology based directly on the conformational changes of H12 in response to different ligands (**Figure 6**). We used a split synthetic Renilla luciferase (RLUC) and Firefly luciferase (FLUC) complementation system, which we previously developed and validated (Paulmurugan and Gambhir 2007; Paulmurugan et al., 2002a), to test this hypothesis by assaying ligand-induced RLUC/FLUC complementation in cell lysates, intact cells, and cell implants in living mice by noninvasive bioluminescence optical imaging. The validated ER intramolecular folding sensors can also be used to distinguish ligand pharmacology in cell culture studies and cell implants in living animals treated with different ER ligands, agonists, selective ER modulators (SERMs), and pure antiestrogens. In adapting this bioluminescence ER ligand sensor system for *in vivo* use, we developed a version that contained a carefully developed single amino acid

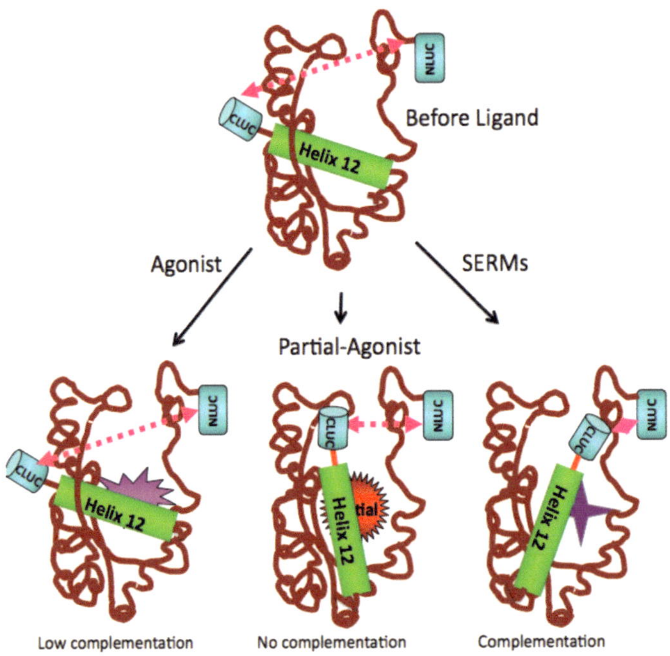

Fig. 6. Schematic representation of the hypothetical model of ligand-induced intramolecular folding of ER that leads to split firefly luciferase (FLUC) complementation. The N- and C-terminal fragments of split-FLUC were fused to the N- and C- terminus, respectively, of the hERα-LBD of various lengths (amino acids 281–549 or 281–595). Binding of ligands to the LBD of ER in the intramolecular folding sensor (N-RLUC-hER-C-RLUC) induces different potential folding patterns in the LBD based on ligands properties of potency and biocharacter. This folding leads to split -FLUC complementation for ER antagonist/SERMS (C), low complementation for ER agonist (A), and no complementation for partial ER agonist/antagonist (B) with the selective folding sensor. Even though the distance between the N- and C-FLUC fragments after binding with partial agonist (B) is smaller than that of agonists (A), this model depicts the importance of the orientations of the split-FLUC fragments in achieving complementation.

mutation in the ERα-LBD, G521T. This ER mutant was selected to be essentially unresponsive to E_2, so that it could be used in mice without interference from the endogenous ligand, while being responsive to certain non-steroidal estrogens such as diethylstilbestrol (DES) and SERMs, making it possible to study their activity *in vivo* by bioluminescence (Paulmurugan and Gambhir 2006; Paulmurugan et al., 2008).

We have developed and validated two hER intramolecular folding sensors that can be used to distinguish ER ligand pharmacology. These receptor sensors can be directly translated from cell culture studies to molecular imaging in small living subjects. In this study we used an ER-based split reporter complementation strategy to follow the position of H12 within the ER–LBD to detect changes in the receptor structural folding in response to ligand binding. The longer construct with the F domain (281–595) appears ligand-pharmacology-independent, whereas the shorter construct without the F domain (281–549) leads to the highest levels of split luciferase complementation in response to SERMs, moderate levels for agonists, and minimal levels for pure antiestrogens (Paulmurugan and Gambhir 2006). We validated these intramolecular folding sensors with various ER ligands in both transiently and stably transfected 293T kidney cells, and MDA-MB-231 (ER-negative) and MCF-7 (ER-positive) breast cancer cells. To extend the folding sensor for applications in living animals, we incorporated a previously undescribed mutant of hER (G521T) into the folding sensor that was insensitive to circulating endogenous estrogen but retained its ability to distinguish SERMs from synthetic agonists. Alternatively, ovariectomized mice can likely be used with the wild-type hER with minimal competition from endogenous estrogens while retaining the ability to study estrogen-like drugs.

To date, several *in vitro* assays have been developed for screening ER ligands by using either purified ERα protein or ER isolated from cell lysates (Inoue et al., 1983; Krey et al., 1997; Nasir and Jolley 1999; Nichols et al., 1998). Limited fluorescence-based assays (Zhou et al., 1998) have been developed to measure receptor conformational changes (23) and recruitment of coactivator peptides (Bai and Giguere 2003; Weatherman et al., 2002; Zhou et al., 1998) in the full-length hERα within cell culture (Michalides et al., 2004). Other assays have been designed to study the effects of synthetic ligands on ER transcription through the activation of downstream target genes (Awais et al., 2004). However, most of these reported assays are not suitable for quantitative, high-throughput screening of ER ligands in intact cells and especially in living subjects through noninvasive molecular imaging. A nontranscriptional assay containing fusion chimeras of either Flp recombinase (Logie et al., 1998) or Cre recombinase (Kemp et al., 2004) with a truncated mouse ERα (amino acids 281–599) has been reported and used for regulating the recombination of reporter genes in cells and living animals. This system demonstrates high background activity even before the addition of ER ligands, mainly through enzymatic amplification, thus limiting its dynamic range in response to different ER ligands. We developed an analogous fusion chimera by fusing a truncated version of hER (amino acids 281–595) with FLUC, which leads to luciferase activity that is 10^4-fold greater than background (mock-transfected cells) even before the addition of ligands. To our knowledge, only one study has reported the construction of mutant versions of hER (G521R and G521V) for selective ER ligand binding using a fusion chimera containing $hER_{251-595}$ with Flp recombinase enzyme (28). Incorporation of the same mutation into our intramolecular folding sensor (N-RLUC-$hER_{281-595}$-C-RLUC) led to nearly complete abolishment of signal for all ER ligands (hER_{G521R}) and a significant reduction in signal (77–89%) for all agonist activities (hER$_{G521V}$) relative to wild type hER. We constructed intramolecular folding sensors using the hER$_{G521}$ mutants with 19 different possible amino acids. We found that the replacement of hER$_{G521}$ with threonine leads to nearly complete abolishment of the E2-induced RLUC complementation but only a 10–20% reduction for all other ER ligands studied. Subsequently, 293T cells stably expressing this intramolecular folding sensor (N-RLUC-hER$_{281-549/G521T}$-C-RLUC) were generated for imaging hERα/ligand complexes in living animals.

The advantages of the intramolecular folding sensor strategy that has been developed and validated include the following: (i) it is real-time (because RLUC exhibits flash kinetics) and quantitative; (ii) it can be used to distinguish binding of agonists, SERMs, and pure antiestrogens; (iii) it can be adapted for studying ligand binding to hER in living animal models by molecular imaging, and thus pharmacokinetic properties of each drug/ligand can be examined; (iv) it allows for a high-throughput strategy for screening/comparing different ER ligands and drugs in multiple cell lines; (v) it allows direct transition from cell culture studies to small living subjects because it is based on a bioluminescence split reporter strategy; and lastly, (vi) it will allow for applications using transgenic models that incorporate the intramolecular folding sensor. In addition, the availability of other split reporters with different properties and substrate specificities should allow multiplexing with other reporter assays.

The limitations with using split RLUC as the reporter gene regarding efflux of its substrate coelenterazine were resolved by showing experiments that resulted in no significant relation between the RLUC complementation and the multidrug resistance systems (Pichler et al., 2004). In addition, the intramolecular folding system was also studied with the improved split FLUC fragments by replacing RLUC fragments. Both systems showed equal sensitivity in different cell culture experiments. The FLUC fragments showed more detectable signal in mouse experiments than RLUC because of more light penetration through tissues, due to the more red-shifted wavelengths of FLUC. Also, the FLUC-based folding system showed greater efficiency in differentiating ER ligands in living mice. It is also possible that the exact locations (cytosolic vs. nuclear) of our fusion reporter proteins may affect the results obtained, and this will need to be explored in future studies. In addition, for some applications *in vivo*, the developed strategies may have difficulty in distinguishing agonists from background, and this potential problem needs to be investigated with testing of additional drugs.

3.5 Reporter protein complementations to monitor multi-protein interactions

The availability of multiple bioluminescent reporters (FLUC, RLUC, GLUC, and possibly VLUC) can be easily adopted for studying multiple proteins involved in a cellular network (e.g., Hsp90 chaperon multi-protein complex has more than 50 proteins) (Goetz et al., 2003) by multiplexing reporter combinations. The use of multiple reporters not only provides the interaction between more than two proteins, but can also provide more precise informations to modulate the effect that one set of protein may exert on other sets of proteins involved in the same complex. In addition, it can also be used to extract the distances among different proteins involved in a complex based on the amount the complementation signals produced from each set of proteins **(Figure 7)**. We are actively exploring multi-protein interactions by multiplexing several combinations, and we hope to soon publish results about this strategy and its feasibility.

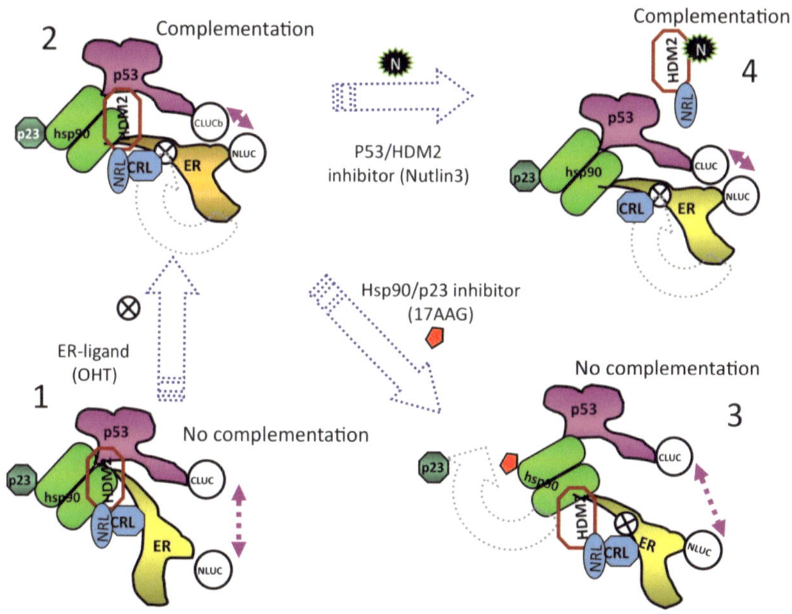

Fig. 7. Schematic representation of two split-reporter complementation systems to detect the interactions between several proteins involved in a multi-protein interaction complex. In this strategy, the results of two reporter protein complementations measured simultaneously between two reporters (Split-FLUC and Split-RLUC) induced by small molecules can predict possible arrangements of different proteins involved in this complex (Hsp90, p23, Estrogen Receptor, p53, and HDM2). Four different hypothetical models have been proposed based on possible theoretical complementations by these reporters in response to drugs which modulate Hsp90/p23 interaction (17AAG), p53/ER interaction (E2), and p53/HDM2 interaction (Nutlin 3).

4. Bioluminescence resonance energy transfer

Bioluminescence Resonance Energy Transfer (BRET) technology involves the nonradioactive transfer of energy between donor and acceptor molecules by the Förster mechanism (46). The energy transfer primarily depends on the following: (1) an overlap between the emission and excitation spectra of the donor (bioluminescence) and acceptor (Fluorophore or a fluorescent protein) molecules, respectively; (2) the proximity of < 100 Å between the donor and the acceptor entities; and (3) the conformational orientation light emission with the acceptance end of the fluorescence entity. As BRET-based technology assumes more prominent roles in the field of studying PPIs, many commercial vendors are developing new instrumentations for measuring BRET ratios, which are generally low-intensity signals. BRET measurements are usually obtained with a microplate reader equipped with specific filter sets for detection of the donor and acceptor emission peaks. This cellular assay has been applied to real-time imaging of cells, high-throughput screening of drugs, and small animal and plant models. There are several combination of BRET involving Renilla luciferase and green fluorescent protein and Firefly luciferase with variants of red fluorescent proteins developed for studying protein-protein interactions. The BRET2 system (Biosignal Packard Montreal, Canada) using renilla luciferase (RLUC) as a bioluminescent donor and mutant GFP2 as a fluorescent acceptor was adapted for expression in mammalian cells and characterized by a significantly red-shifted Stokes shift that emits transferred energy at 508 nm. The resonance energy transfer from the reaction of the reconstructed RLUC protein with its substrate Deep Blue Coelenterazine (DBC) excites the GFP2 protein, as the two fused proteins Id and MyoD, or FKBP12 and FRB which interact in the presence of a small molecule mediator (rapamycin). Our lab also demonstrated the ability to detect signal from PPIs in cultured cells, as well as from the surface and deeper tissues of small living animals with implanted cells over expressing the fusion constructs (For further details, read (De et al., 2009; Dragulescu-Andrasi et al., 2011).

Our lab has recently showed that the BRET2 assay sensitivity can be significantly improved by using RLUC mutants with improved quantum efficiency and/or stability (eg, RLUC8 and RLUCM) as a donor. To extend the time of light measurement, we also developed CLZ400 (also known as bisdeoxycoelenterazine) analogs, showing that signal from our improved BRET2 vector can be monitored for up to 6 hours. This approach, currently undergoing continuing validation, should have important implications for the study of PPIs in cells maintained in their natural environment, particularly if it can be effectively applied for the evaluation of new pharmaceuticals. Most recently, further advances in this field have led us to develop a high photon efficiency, self-illuminating fusion protein combining a mutant red fluorescent protein (mOrange) and a mutant RLUC (RLUC8). This new BRET fusion protein (BRET3) exhibits a several fold improvement in light intensity in comparison to existing BRET fusion proteins. BRET3 also exhibits the most red-shifted light output (564 nm peak wavelength) of any reported bioluminescence protein that uses its natural coelenterazine substrate, a benefit that can be demonstrated at various tissue depths in small animals.

5. Future directions in bioluminescence imaging

Molecular imaging has been recognized as an important and exciting area of bio-medical research, mainly because of its ability to visually represent, characterize, and quantify biological processes in living subjects. Techniques such as Positron Emission Tomography (PET), Single-photon Emission Computed Tomography (SPECT), and Magnetic Resonance Imaging (MRI) have been extensively used in the clinic for several diagnostic and disease monitoring processes; all these systems explore intracellular proteins or other molecules as probes for the signal. Reporter genes (Bioluminescence, fluorescence, and PET), on the other hand, are capable of precisely monitoring sub-cellular processes and their native functional actions in cells, and imaging them in living animals. Challenges, however, remain in delivering these proteins in cells without perturbing the cellular microenvironments. Another obstacle is in generating sufficient sensitivity to measure theses signals, especially in living animals. Problems associated with the modulations in the cellular microenvironment are also tricky, but may be minimized by expressing few copies and weighing the sensitivity. The continuing development of new high-sensitivity instruments with tomographic imaging capabilities and improved spatial resolutions will play an important role in expanding the applications of bioluminescent reporters and exploiting their unique ability to precisely image the sub-cellular processes in their native microenvironment.

6. References

Anfinsen CB. (1973). Principles that govern the folding of protein chains. *Science Vol.181*, No.96:pp.223-230.

Artemov D, Mori N, Ravi R, and Bhujwalla ZM. (2003). Magnetic resonance molecular imaging of the HER-2/neu receptor. *Cancer Res Vol.63*, No.11:pp.2723-2727.

Awais M, Sato M, Sasaki K, and Umezawa Y. (2004). A genetically encoded fluorescent indicator capable of discriminating estrogen agonists from antagonists in living cells. *Anal Chem Vol.76*, No.8:pp.2181-2186.

Bai Y, and Giguere V. (2003). Isoform-selective interactions between estrogen receptors and steroid receptor coactivators promoted by estradiol and ErbB-2 signaling in living cells. *Mol Endocrinol Vol.17*, No.4:pp.589-599.

Beck V, Rohr U, and Jungbauer A. (2005). Phytoestrogens derived from red clover: an alternative to estrogen replacement therapy? *J Steroid Biochem Mol Biol Vol.94*, No.5:pp.499-518.

Becker T, Weber K, and Johnsson N. (1990). Protein-protein recognition via short amphiphilic helices; a mutational analysis of the binding site of annexin II for p11. *Embo J Vol.9*, No.13:pp.4207-4213.

Beeckmans S. (1999). Chromatographic methods to study protein-protein interactions. *Methods Vol.19*, No.2:pp.278-305.

Bhaumik S, and Gambhir SS. (2002a). Optical imaging of Renilla luciferase reporter gene expression in living mice. *Proc Natl Acad Sci U S A Vol.99*, No.1:pp.377-382.

Bhaumik S, and Gambhir SS. (2002b). Optical imaging of Renilla luciferase reporter gene expression in living mice. *Proc Natl Acad Sci USA Vol.99*, No.1:pp.377-382.

Bhaumik S, Lewis XZ, and Gambhir SS. (2004). Optical imaging of Renilla luciferase, synthetic Renilla luciferase, and firefly luciferase reporter gene expression in living mice. *J Biomed Opt Vol.9*, No.3:pp.578-586.

Bode J, and Willmitzer. (1975). Application of fluorescamine to the study of protein-DNA interactions. *Nucleic Acids Res Vol.2*, No.10:pp.1951-1965.

Brzozowski AM, Pike AC, Dauter Z, Hubbard RE, Bonn T, Engstrom O, Ohman L, Greene GL, Gustafsson JA, and Carlquist M. (1997). Molecular basis of agonism and antagonism in the oestrogen receptor. *Nature Vol.389*, No.6652:pp.753-758.

Burt BM, Humm JL, Kooby DA, Squire OD, Mastorides S, Larson SM, and Fong Y. (2001). Using positron emission tomography with [(18)F]FDG to predict tumor behavior in experimental colorectal cancer. *Neoplasia Vol.3*, No.3:pp.189-195.

Cherry SR, and Gambhir SS. (2001). Use of positron emission tomography in animal research. *Ilar J Vol.42*, No.3:pp.219-232.

Contag CH, Jenkins D, Contag PR, and Negrin RS. (2000). Use of reporter genes for optical measurements of neoplastic disease in vivo. *Neoplasia Vol.2*, No.1-2:pp.41-52.

Contag CH, and Ross BD. (2002). It's not just about anatomy: in vivo bioluminescence imaging as an eyepiece into biology. *J Magn Reson Imaging Vol.16*, No.4:pp.378-387.

Contag CH, Spilman SD, Contag PR, Oshiro M, Eames B, Dennery P, Stevenson DK, and Benaron DA. (1997). Visualizing gene expression in living mammals using a bioluminescent reporter. *Photochem Photobiol Vol.66*, No.4:pp.523-531.

Contag PR, Olomu IN, Stevenson DK, and Contag CH. (1998). Bioluminescent indicators in living mammals. *Nat Med Vol.4*, No.2:pp.245-247.

De A, Ray P, Loening AM, and Gambhir SS. (2009). BRET3: a red-shifted bioluminescence resonance energy transfer (BRET)-based integrated platform for imaging protein-protein interactions from single live cells and living animals. *FASEB J Vol.23*, No.8:pp.2702-2709.

de Wet JR, Wood KV, Helinski DR, and DeLuca M. (1985). Cloning of firefly luciferase cDNA and the expression of active luciferase in Escherichia coli. *Proc Natl Acad Sci U S A Vol.82*, No.23:pp.7870-7873.

Dragulescu-Andrasi A, Chan CT, De A, Massoud TF, and Gambhir SS. (2011). Bioluminescence resonance energy transfer (BRET) imaging of protein-protein interactions within deep tissues of living subjects. *Proc Natl Acad Sci U S A Vol.108,* No.29:pp.12060-12065.

Fernandez-Guinea O, Andicoechea A, Gonzalez LO, Gonzalez-Reyes S, Merino AM, Hernandez LC, Lopez-Muniz A, Garcia-Pravia P, and Vizoso FJ. (2010). Relationship between morphological features and kinetic patterns of enhancement of the dynamic breast magnetic resonance imaging and clinico-pathological and biological factors in invasive breast cancer. *BMC Cancer Vol.10*:pp.8.

Fields S, and Song O. (1989). A novel genetic system to detect protein-protein interactions. *Nature Vol.340,* No.6230:pp.245-246.

Gambhir SS. (2002). Molecular imaging of cancer with positron emission tomography. *Nat Rev Cancer Vol.2,* No.9:pp.683-693.

Goetz MP, Toft DO, Ames MM, and Erlichman C. (2003). The Hsp90 chaperone complex as a novel target for cancer therapy. *Ann Oncol Vol.14,* No.8:pp.1169-1176.

Goldenberg DM, Preston DF, Primus FJ, and Hansen HJ. (1974). Photoscan localization of GW-39 tumors in hamsters using radiolabeled anticarcinoembryonic antigen immunoglobulin G. *Cancer Res Vol.34,* No.1:pp.1-9.

Goldenberg DM, and Sharkey RM. (2007). Novel radiolabeled antibody conjugates. *Oncogene Vol.26,* No.25:pp.3734-3744.

Gronemeyer H. (1991). Transcription activation by estrogen and progesterone receptors. *Annu Rev Genet Vol.25*:pp.89-123.

Hastings JW. (1996). Chemistries and colors of bioluminescent reactions: a review. *Gene Vol.173,* No.1 Spec No:pp.5-11.

Hodges RA, Perler FB, Noren CJ, and Jack WE. (1992). Protein splicing removes intervening sequences in an archaea DNA polymerase. *Nucleic Acids Res Vol.20,* No.23:pp.6153-6157.

Hucho F, and Wallenfels K. (1972). Glucono- -lactonase from Escherichia coli. *Biochim Biophys Acta Vol.276,* No.1:pp.176-179.

Inoue A, Yamakawa J, Yukioka M, and Morisawa S. (1983). Filter-binding assay procedure for thyroid hormone receptors. *Anal Biochem Vol.134,* No.1:pp.176-183.

Jawhara S, and Mordon S. (2004). In vivo imaging of bioluminescent Escherichia coli in a cutaneous wound infection model for evaluation of an antibiotic therapy. *Antimicrob Agents Chemother Vol.48,* No.9:pp.3436-3441.

Johnsson N, and Varshavsky A. (1994a). Split ubiquitin as a sensor of protein interactions in vivo. *Proc Natl Acad Sci U S A Vol.91,* No.22:pp.10340-10344.

Johnsson N, and Varshavsky A. (1994b). Ubiquitin-assisted dissection of protein transport across membranes. *Embo J Vol.13,* No.11:pp.2686-2698.

Jung JC, and Schnitzer MJ. (2003). Multiphoton endoscopy. *Opt Lett Vol.28,* No.11:pp.902-904.

Kemp R, Ireland H, Clayton E, Houghton C, Howard L, and Winton DJ. (2004). Elimination of background recombination: somatic induction of Cre by combined transcriptional regulation and hormone binding affinity. *Nucleic Acids Res Vol.32,* No.11:pp.e92.

Kippen AD, and Fersht AR. (1995). Analysis of the mechanism of assembly of cleaved barnase from two peptide fragments and its relevance to the folding pathway of uncleaved barnase. *Biochemistry Vol.34,* No.4:pp.1464-1468.

Krey G, Braissant O, L'Horset F, Kalkhoven E, Perroud M, Parker MG, and Wahli W. (1997). Fatty acids, eicosanoids, and hypolipidemic agents identified as ligands of peroxisome proliferator-activated receptors by coactivator-dependent receptor ligand assay. *Mol Endocrinol Vol.11*, No.6:pp.779-791.

Levinthal C. (1969). "How to Fold Graciously". Mossbauer Spectroscopy in Biological Systems: . *Proceedings of a meeting held at Allerton House, Monticello, Illinois: 22–24.*

Logie C, Nichols M, Myles K, Funder JW, and Stewart AF. (1998). Positive and negative discrimination of estrogen receptor agonists and antagonists using site-specific DNA recombinase fusion proteins. *Mol Endocrinol Vol.12*, No.8:pp.1120-1132.

Loontiens FG, Wallenfels K, Weil R, and Massart EL. (1970). [Interaction between "Escherichia coli" beta-galactosidase and O-mercuriphenyl beta-D-galactoside chloride, an unexpected substrate]. *Arch Int Physiol Biochim Vol.78*, No.3:pp.596-597.

Lorenz WW, Cormier MJ, O'Kane DJ, Hua D, Escher AA, and Szalay AA. (1996). Expression of the Renilla reniformis luciferase gene in mammalian cells. *J Biolumin Chemilumin Vol.11*, No.1:pp.31-37.

Lorenz WW, McCann RO, Longiaru M, and Cormier MJ. (1991). Isolation and expression of a cDNA encoding Renilla reniformis luciferase. *Proc Natl Acad Sci U S A Vol.88*, No.10:pp.4438-4442.

Luker GD, Sharma V, Pica CM, Dahlheimer JL, Li W, Ochesky J, Ryan CE, Piwnica-Worms H, and Piwnica-Worms D. (2002a). Noninvasive imaging of protein-protein interactions in living animals. *Proc Natl Acad Sci U S A Vol.99*, No.10:pp.6961-6966.

Luker GD, Sharma V, Pica CM, Dahlheimer JL, Li W, Ochesky J, Ryan CE, Piwnica-Worms H, and Piwnica-Worms D. (2002b). Noninvasive imaging of protein-protein interactions in living animals. *Proc Natl Acad Sci U S A Vol.99*, No.10:pp.6961-6966.

Luker GD, Sharma V, Pica CM, Prior JL, Li W, and Piwnica-Worms D. (2003). Molecular imaging of protein-protein interactions: controlled expression of p53 and large T-antigen fusion proteins in vivo. *Cancer Res Vol.63*, No.8:pp.1780-1788.

Luker KE, Smith MC, Luker GD, Gammon ST, Piwnica-Worms H, and Piwnica-Worms D. (2004). Kinetics of regulated protein-protein interactions revealed with firefly luciferase complementation imaging in cells and living animals. *Proc Natl Acad Sci U S A Vol.101*, No.33:pp.12288-12293.

Luo Y, Batalao A, Zhou H, and Zhu L. (1997). Mammalian two-hybrid system: a complementary approach to the yeast two-hybrid system. *Biotechniques Vol.22*, No.2:pp.350-352.

MacDonald SM, Harisinghani MG, Katkar A, Napolitano B, Wolfgang J, and Taghian AG. (2010). Nanoparticle-enhanced MRI to evaluate radiation delivery to the regional lymphatics for patients with breast cancer. *Int J Radiat Oncol Biol Phys Vol.77*, No.4:pp.1098-1104.

Massoud TF, and Gambhir SS. (2003). Molecular imaging in living subjects: seeing fundamental biological processes in a new light. *Genes Dev Vol.17*, No.5:pp.545-580.

Massoud TF, Paulmurugan R, and Gambhir SS. (2004). Molecular imaging of homodimeric protein-protein interactions in living subjects. *Faseb J Vol.18*, No.10:pp.1105-1107.

Massoud TF, Paulmurugan R, and Gambhir SS. (2010). A molecularly engineered split reporter for imaging protein-protein interactions with positron emission tomography. *Nat Med Vol.16*, No.8:pp.921-926.

Mehta AD, Jung JC, Flusberg BA, and Schnitzer MJ. (2004). Fiber optic in vivo imaging in the mammalian nervous system. *Curr Opin Neurobiol Vol.14*, No.5:pp.617-628.

Michalides R, Griekspoor A, Balkenende A, Verwoerd D, Janssen L, Jalink K, Floore A, Velds A, van't Veer L, and Neefjes J. (2004). Tamoxifen resistance by a conformational arrest of the estrogen receptor alpha after PKA activation in breast cancer. *Cancer Cell Vol.5*, No.6:pp.597-605.

Michnick SW. (2001). Exploring protein interactions by interaction-induced folding of proteins from complementary peptide fragments. *Curr Opin Struct Biol Vol.11*, No.4:pp.472-477.

Mintun MA, Welch MJ, Siegel BA, Mathias CJ, Brodack JW, McGuire AH, and Katzenellenbogen JA. (1988). Breast cancer: PET imaging of estrogen receptors. *Radiology Vol.169*, No.1:pp.45-48.

Morris SJ, Sudhof TC, and Haynes DH. (1982). Calcium-promoted resonance energy transfer between fluorescently labeled proteins during aggregation of chromaffin granule membranes. *Biochim Biophys Acta Vol.693*, No.2:pp.425-436.

Nasir MS, and Jolley ME. (1999). Fluorescence polarization: an analytical tool for immunoassay and drug discovery. *Comb Chem High Throughput Screen Vol.2*, No.4:pp.177-190.

Nichols JS, Parks DJ, Consler TG, and Blanchard SG. (1998). Development of a scintillation proximity assay for peroxisome proliferator-activated receptor gamma ligand binding domain. *Anal Biochem Vol.257*, No.2:pp.112-119.

Nichtl A, Buchner J, Jaenicke R, Rudolph R, and Scheibel T. (1998). Folding and association of beta-Galactosidase. *J Mol Biol Vol.282*, No.5:pp.1083-1091.

Ntziachristos V, Tung CH, Bremer C, and Weissleder R. (2002). Fluorescence molecular tomography resolves protease activity in vivo. *Nat Med Vol.8*, No.7:pp.757-760.

Ozawa T, Kaihara A, Sato M, Tachihara K, and Umezawa Y. (2001). Split luciferase as an optical probe for detecting protein-protein interactions in mammalian cells based on protein splicing. *Anal Chem Vol.73*, No.11:pp.2516-2521.

Paulmurugan R, and Gambhir SS. (2003). Monitoring protein-protein interactions using split synthetic renilla luciferase protein-fragment-assisted complementation. *Anal Chem Vol.75*, No.7:pp.1584-1589.

Paulmurugan R, and Gambhir SS. (2006). An intramolecular folding sensor for imaging estrogen receptor-ligand interactions. *Proc Natl Acad Sci U S A Vol.103*, No.43:pp.15883-15888.

Paulmurugan R, and Gambhir SS. (2007). Combinatorial library screening for developing an improved split-firefly luciferase fragment-assisted complementation system for studying protein-protein interactions. *Anal Chem Vol.79*, No.6:pp.2346-2353.

Paulmurugan R, Massoud TF, Huang J, and Gambhir SS. (2004). Molecular imaging of drug-modulated protein-protein interactions in living subjects. *Cancer Res Vol.64*, No.6:pp.2113-2119.

Paulmurugan R, Padmanabhan P, Ahn BC, Ray S, Willmann JK, Massoud TF, Biswal S, and Gambhir SS. (2009). A novel estrogen receptor intramolecular folding-based titratable transgene expression system. *Mol Ther Vol.17*, No.10:pp.1703-1711.

Paulmurugan R, Tamrazi A, Katzenellenbogen JA, Katzenellenbogen BS, and Gambhir SS. (2008). A human estrogen receptor (ER)alpha mutation with differential responsiveness to nonsteroidal ligands: novel approaches for studying mechanism of ER action. *Mol Endocrinol Vol.22*, No.7:pp.1552-1564.

Paulmurugan R, Umezawa Y, and Gambhir SS. (2002a). Noninvasive imaging of protein-protein interactions in living subjects by using reporter protein complementation and reconstitution strategies. *Proc Natl Acad Sci U S A Vol.99*, No.24:pp.15608-15613.

Paulmurugan R, Umezawa Y, and Gambhir SS. (2002b). Noninvasive Imaging of Protein-Protein Interactions in Living Subjects Using Reporter Protein Complementation and Reconstitution Strategies. *Proc Natl Acad Sci U S A Vol.99*, No.24:pp.15608-15613.

Pelletier JN, Arndt KM, Pluckthun A, and Michnick SW. (1999). An in vivo library-versus-library selection of optimized protein-protein interactions. *Nat Biotechnol Vol.17*, No.7:pp.683-690.

Pichler A, Prior JL, and Piwnica-Worms D. (2004). Imaging reversal of multidrug resistance in living mice with bioluminescence: MDR1 P-glycoprotein transports coelenterazine. *Proc Natl Acad Sci U S A Vol.101*, No.6:pp.1702-1707.

Pike AC, Brzozowski AM, Walton J, Hubbard RE, Bonn T, Gustafsson JA, and Carlquist M. (2000). Structural aspects of agonism and antagonism in the oestrogen receptor. *Biochem Soc Trans Vol.28*, No.4:pp.396-400.

Ray P, Pimenta H, Paulmurugan R, Berger F, Phelps ME, Iyer M, and Gambhir SS. (2002a). Noninvasive quantitative imaging of protein-protein interactions in living subjects. *Proc Natl Acad Sci U S A Vol.99*, No.5:pp.3105-3110.

Ray P, Pimenta H, Paulmurugan R, Berger F, Phelps ME, Iyer M, and Gambhir SS. (2002b). Noninvasive quantitative imaging of protein-protein interactions in living subjects. *Proc Natl Acad Sci U S A Vol.99*, No.5:pp.3105-3110.

Ray P, Wu AM, and Gambhir SS. (2003). Optical bioluminescence and positron emission tomography imaging of a novel fusion reporter gene in tumor xenografts of living mice. *Cancer Res Vol.63*, No.6:pp.1160-1165.

Remy I, and Michnick SW. (1999). Clonal selection and in vivo quantitation of protein interactions with protein-fragment complementation assays. *Proc Natl Acad Sci U S A Vol.96*, No.10:pp.5394-5399.

Rossi F, Charlton CA, and Blau HM. (1997). Monitoring protein-protein interactions in intact eukaryotic cells by beta-galactosidase complementation. *Proc Natl Acad Sci U S A Vol.94*, No.16:pp.8405-8410.

Russwurm M, Mullershausen F, Friebe A, Jager R, Russwurm C, and Koesling D. (2007). Design of fluorescence resonance energy transfer (FRET)-based cGMP indicators: a systematic approach. *Biochem J Vol.407*, No.1:pp.69-77.

Schuur ER, Loktev AV, Sharma M, Sun Z, Roth RA, and Weigel RJ. (2001). Ligand-dependent interaction of estrogen receptor-alpha with members of the forkhead transcription factor family. *J Biol Chem Vol.276*, No.36:pp.33554-33560.

Scopinaro F, Varvarigou AD, Ussof W, De Vincentis G, Sourlingas TG, Evangelatos GP, Datsteris J, and Archimandritis SC. (2002). Technetium labeled bombesin-like peptide: preliminary report on breast cancer uptake in patients. *Cancer Biother Radiopharm Vol.17*, No.3:pp.327-335.

Shiau AK, Barstad D, Loria PM, Cheng L, Kushner PJ, Agard DA, and Greene GL. (1998). The structural basis of estrogen receptor/coactivator recognition and the antagonism of this interaction by tamoxifen. *Cell Vol.95*, No.7:pp.927-937.

Skolnick J, Fetrow JS, and Kolinski A. (2000). Structural genomics and its importance for gene function analysis. *Nat Biotechnol Vol.18*, No.3:pp.283-287.

Smith VF, and Matthews CR. (2001). Testing the role of chain connectivity on the stability and structure of dihydrofolate reductase from E. coli: fragment complementation and circular permutation reveal stable, alternatively folded forms. *Protein Sci Vol.10*, No.1:pp.116-128.

Stagljar I, Korostensky C, Johnsson N, and te Heesen S. (1998). A genetic system based on split-ubiquitin for the analysis of interactions between membrane proteins in vivo. *Proc Natl Acad Sci U S A Vol.95*, No.9:pp.5187-5192.

Stopeck AT, Hersh EM, Brailey JL, Clark PR, Norman J, and Parker SE. (1998). Transfection of primary tumor cells and tumor cell lines with plasmid DNA/lipid complexes. *Cancer Gene Ther Vol.5*, No.2:pp.119-126.

Szabo BK, Aspelin P, Kristoffersen Wiberg M, Tot T, and Bone B. (2003). Invasive breast cancer: correlation of dynamic MR features with prognostic factors. *Eur Radiol Vol.13*, No.11:pp.2425-2435.

Tang Y, Wang J, Scollard DA, Mondal H, Holloway C, Kahn HJ, and Reilly RM. (2005). Imaging of HER2/neu-positive BT-474 human breast cancer xenografts in athymic mice using (111)In-trastuzumab (Herceptin) Fab fragments. *Nucl Med Biol Vol.32*, No.1:pp.51-58.

Ter-Pogossian MM, Phelps ME, Hoffman EJ, and Mullani NA. (1975). A positron-emission transaxial tomograph for nuclear imaging (PETT). *Radiology Vol.114*, No.1:pp.89-98.

Torigian DA, Huang SS, Houseni M, and Alavi A. (2007). Functional imaging of cancer with emphasis on molecular techniques. *CA Cancer J Clin Vol.57*, No.4:pp.206-224.

Tsai MJ, and O'Malley BW. (1994). Molecular mechanisms of action of steroid/thyroid receptor superfamily members. *Annu Rev Biochem Vol.63*:pp.451-486.

Tsien RY. (2009). Indicators based on fluorescence resonance energy transfer (FRET). *Cold Spring Harb Protoc Vol.2009*, No.7:pp.pdb top57.

Tucker CL, Gera JF, and Uetz P. (2001). Towards an understanding of complex protein networks. *Trends Cell Biol Vol.11*, No.3:pp.102-106.

Tung CH, Bredow S, Mahmood U, and Weissleder R. (1999). Preparation of a cathepsin D sensitive near-infrared fluorescence probe for imaging. *Bioconjug Chem Vol.10*, No.5:pp.892-896.

Ullmann A, Jacob F, and Monod J. (1968). On the subunit structure of wild-type versus complemented beta-galactosidase of Escherichia coli. *J Mol Biol Vol.32*, No.1:pp.1-13.

Weatherman RV, Chang CY, Clegg NJ, Carroll DC, Day RN, Baxter JD, McDonnell DP, Scanlan TS, and Schaufele F. (2002). Ligand-selective interactions of ER detected in living cells by fluorescence resonance energy transfer. *Mol Endocrinol Vol.16*, No.3:pp.487-496.

Wehrman T, Kleaveland B, Her JH, Balint RF, and Blau HM. (2002). Protein-protein interactions monitored in mammalian cells via complementation of beta -lactamase enzyme fragments. *Proc Natl Acad Sci U S A Vol.99*, No.6:pp.3469-3474.

Weissleder R. (2002). Scaling down imaging: molecular mapping of cancer in mice. *Nat Rev Cancer Vol.2*, No.1:pp.11-18.

Weissleder R, and Pittet MJ. (2008). Imaging in the era of molecular oncology. *Nature Vol.452*, No.7187:pp.580-589.

Wills PR. (2001). Assessing the Effects - Protein Interaction Networks. *http://wwwphyaucklandacnz/staff/prw/biocomplexity/protein_networkhtm.*

Wilson T, and Hastings JW. (1998). Bioluminescence. *Annu Rev Cell Dev Biol Vol.14*:pp.197-230.

Wu JC, Inubushi M, Sundaresan G, Schelbert HR, and Gambhir SS. (2002). Optical imaging of cardiac reporter gene expression in living rats. *Circulation Vol.105*, No.14:pp.1631-1634.

Wu JC, Sundaresan G, Iyer M, and Gambhir SS. (2001). Noninvasive optical imaging of firefly luciferase reporter gene expression in skeletal muscles of living mice. *Mol Ther Vol.4*, No.4:pp.297-306.

Zhou G, Cummings R, Li Y, Mitra S, Wilkinson HA, Elbrecht A, Hermes JD, Schaeffer JM, Smith RG, and Moller DE. (1998). Nuclear receptors have distinct affinities for coactivators: characterization by fluorescence resonance energy transfer. *Mol Endocrinol Vol.12*, No.10:pp.1594-1604.

4

Quantitative Assessment of Seven Transmembrane Receptors (7TMRs) Oligomerization by Bioluminescence Resonance Energy Transfer (BRET) Technology

Valentina Kubale[1], Luka Drinovec[2] and Milka Vrecl[1]
[1]*Institute of Anatomy, Histology & Embryology, Veterinary Faculty of University in Ljubljana,*
[2]*Aerosol d.o.o., Ljubljana,*
Slovenia

1. Introduction

Seven transmembrane receptors (7TMRs; also designated as G-protein coupled receptors (GPCRs)) form the largest and evolutionarily well conserved family of cell-surface receptors, with more than 800 members identified in the human genome. 7TMRs are the targets both for a plethora of endogenous ligands (e.g. peptides, glycoproteins, lipids, amino acids, nucleotides, neurotransmitters, odorants, ions, and photons) and therapeutic drugs and transduce extracellular stimuli into intracellular responses mainly via coupling to guanine nucleotide binding proteins (G-proteins) (McGraw & Liggett, 2006).

These receptors have traditionally been viewed as monomeric entities and only more recent biochemical and biophysical studies have changed this view. The idea that 7TMRs might form dimers or higher order oligomeric complexes has been formulated more than 20 years ago and since then intensively studied. In the last decade, bioluminescence resonance energy transfer (BRET) was one of the most commonly used biophysical methods to study 7TMRs oligomerization. This technique enables monitoring physical interactions between protein partners in living cells fused to donor and acceptor moieties. It relies on non-radiative transfer of energy between donor and acceptor, their intermolecular distance (10 - 100 Å) and relative orientation. Over this period the method has progressed and several versions of BRET have been developed that use different substrates and/or energy donor/acceptor couples to improve stability and specificity of the BRET signal. This chapter outlines BRET-based approaches to study 7TMRs oligomerization (e.g. BRET saturation and competition assays), control experiments needed in the interpretation i.e. establishing specificity of BRET results and mathematical models applied to quantitatively assess the oligomerization state of studied receptors.

2. Seven transmembrane receptors (7TMRs): Structure and characteristics

Primary sequence comparisons reveal that 7TMRs share sequence and topology similarities allowing them to be classified as a super-gene family. These receptors are characterized by seven hydrophobic stretches of 20-25 amino acids, predicted to form transmembrane α-helices. Prediction of transmembrane folding was based largely on the method proposed by Kyte and Doolitle (Kyte & Doolittle, 1982). This method plots the hydrophobicity of the amino acids along the sequence, assigning each amino acids a hydrophobicity index. By summing this index over a window of nine residues, the transmembrane sequence is postulated when index reaches the value of 1.6 for a stretch of ~20 amino acids. This number is based on the assumption that the membrane spanning sequences of protein are α-helical and that about six helical turns are required to span the lipid bilayer (Hucho & Tsetlin, 1996). The highly hydrophobic α-helices that serve as transmembrane spanning domains (TMs) are connected by three extracellular (ECL) and three intracellular (ICL) hydrophilic loops. Amino (N)-terminal fragment is extracellular and the carboxyl (C)-terminal tail is intracellular. In the recent years this common structural topology was also confirmed by three-dimensional crystal structure of some 7TMR members (reviewed by (Salon et al., 2011)). Additionally, 7TMRs may undergo a variety of posttranslational modifications such as N-linked glycosylation, formation of disulfide bonds, palmitoylation and phosphorylation. 7TMRs contain at least one consensus sequence for N-linked glycosylation (Asn-x-Ser/Thr), usually located near the N-terminus, although there are potential glycosylation sites in the intracellular loops. They also contain a number of conserved extracellular cysteine residues, some of which appear to play a role in stabilizing the receptor's tertiary structure. An additional highly conserved cysteine can be present within the C-terminal tail of many 7TMRs. When palmitoylated, it may anchor a part of cytoplasmic tail of the receptor to the plasma membrane, thus forming the fourth ICL and controlling the tertiary structure. Consensus sequences for potential phosphorylation sites (serine and threonine residues) are located in the second and third ICLs, and in particular, in the intracellular C-terminal tail. The most obvious structural differences between the receptors in subgroups are the length of their N-terminal fragment and the loops between TMs. Originally, 7TMRs were divided into six groups, A – F; families (also known as "groups" or "classes") A, B and C included all mammalian 7TMRs. Genome projects then generated numerous new 7TM sequences and more than 800 human 7TMRs were reclassified into five families, A – E (reviewed by (Gurevich & Gurevich, 2006; Salon et al., 2011)).

Family A (also known as the rhodopsin family) is by far the largest family of 7TMRs (containing ~700 members), and includes many of the receptors for biogenic amines and small peptides. It is characterized by very short N- and C-termini as well as several highly conserved amino acids. In most cases TMs serve as the ligand-binding site. This family contains some of the most extensively studied 7TMRs, the opsins and the β-adrenergic receptors. Recent structural information for a few family A 7TMR members (e.g. rhodopsin, opsin, human β_2-adrenergic receptor, turkey β_1-adrenergic receptor, human A_{2A}-adenosine receptor, CXC chemokine receptor type 4 and D_3-dopamine receptor) confirmed an obvious conservation of the topology and seven-transmembrane architecture (Salon et al., 2011). Family B (secretin-receptor family), which has considerably fewer members i.e. 15, is characterized by a long N-terminus (>400 amino acids) containing six conserved cysteine residues that contribute to three conserved disulfide bonds, which provide structural stability, and a conserved cleft for the docking of often helical C-terminal region of the peptide ligands. Natural ligands for family B 7TMRs are all moderately large peptides, such as calcitonin, parathyroid hormone and glucagon. Family C (metabotropic glutamate family) contains 15 members that are the metabotropic glutamate receptors (mGluRs), the

Ca^{2+} sensing receptor, and the receptor for the major excitatory neurotransmitter in the central nervous system, the γ-aminobutyric acid ($GABA_B$) receptor and orphan receptors. This family has a very large N-terminal domain (>600 amino acids), which bears the agonist binding site and also a long C-tail (Kenakin & Miller, 2010; McGraw & Liggett, 2006). Notably, family C members form obligatory dimers (Kniazeff et al., 2011). Two ancillary families consist of class D (adhesion family), containing 24 members, and class E (frizzled family), with 24 members.

3. 7TMRs homo- and hetero-oligomerization

In 1983, Fuxe et al. (Fuxe et al., 1983) formulated the hypothesis about the existence of homo-dimers for different types of 7TMRs and in the same year the first demonstration of 7TMRs homo-dimers and homo-tetramers of muscarinic receptors was published (Avissar et al., 1983). However, the evidence for dimerization existed even before that. Following classical radio-ligand studies on the insulin receptor (de Meyts et al., 1973), negative cooperativity, for which dimerization is a prerequisite, has also been demonstrated for β_2-adrenergic receptor (β_2-AR) (Limbird et al., 1975) and thyrotrophin-stimulating hormone (TSH) receptor (De Meyts, 1976) binding in the early 70's, before they were shown to be 7TMRs and this issue remained controversial for over two decades. 7TMRs can be either connected to identical partner(s), which results in formation of homo-dimers (or homo-oligomers), or to structurally different receptor(s), which results in formation of hetero-dimers (hetero-oligomers). 7TMR dimerization was proposed to play a potential role in i) receptor maturation and correct transport to the plasma membrane, ii) ligand-promoted regulation, iii) pharmacological diversity (e.g. positive and negative ligand binding cooperativity), iv) signal transduction (potentiating/attenuating signaling or changing G-protein selectivity), and v) receptor internalization and desensitization (Terrillon & Bouvier, 2004). The first widely accepted demonstration of 7TMR hetero-dimerization came from the $GABA_B$ (GBBR) receptors that exclusively function in a heteromeric form (White et al., 1998).

There is now considerable evidence to indicate that 7TMRs can form and function as homo-dimers and hetero-dimers (reviewed by (Filizola, 2010; Gurevich & Gurevich, 2008a; Palczewski, 2010)) and that these dimers may have therapeutic relevance (Casado et al., 2009). Hetero-dimerization in the C family of receptors has been most extensively studied and for some experts in the field of 7TMRs the only one demonstrated to form real dimers (for recent review see (Kniazeff et al., 2011)). In this family of 7TMRs receptors hetero-dimerization is important for either receptor function, proper expression on the cell surface or enhancing receptor activity. In the most numerous family A 7TMRs dimerization was extensively studied, although with few exceptions functional role of receptor self-association is in most cases unclear. Compelling evidence for the dimerization in the family A 7TMR was only recently demonstrated *in vivo* by Huhtaniemi's group, who was able to rescue the LH receptor knockout phenotype by complementation i.e. co-expressing two nonfunctional receptor mutants in the knockout mice (Rivero-Muller et al., 2010). Members of the family B 7TMRs have also only recently been shown to associate as stable homo-dimers. The structural basis of this, at least for the prototypic secretin receptor, is the lipid-exposed face of TM4. This complex has been postulated as being important for the structural stabilization of the high affinity complex with G-protein (reviewed by (reviewed by (Kenakin & Miller, 2010)).

In addition to widespread intra-family hetero-dimerization, inter-family hetero-dimerization has also been reported, at least between both of the family A members β_2-AR and opsin and the family B member gastric inhibitory polypeptide receptor (GIP) (Vrecl et al., 2006), and between the family A serotonin 5-HT_{2A} receptors and the family C mGluR2 (Gonzalez-Maeso et al., 2008). Both types of hetero-dimers were demonstrated to be functional, either by their ability to induce cAMP production upon agonist stimulation (family A/B hetero-dimer), or by their ability to modulate G-protein coupling (family A/C hetero-dimer).

3.1 Dimerization interface

Growing experimental data support the view that 7TMRs exist and function as contact dimers or higher order oligomers with TM regions at the interfaces. In contact dimers/oligomers of 7TMRs, the original TM helical-bundle topology of each individual protomer is preserved and interaction interfaces are formed by lipid-exposed surfaces. Although domain-swap models, i.e. models in which domains TM1/TM5 and TM6/TM7 would exchange between protomers, have also been proposed in the literature, there is there is limited direct evidence that supports these assumptions. On the other hand, compelling experimental evidence exists for the involvement of lipid exposed surfaces of TM1, TM4 and/or TM5 at the dimerization/oligomerization interfaces of several 7TMRs. Besides, the interface may depend on additional stabilizing factors such as the coiled-coil interactions reported in the $GABA_B$ receptor and the disulfide bridge interactions in the muscarinic and the other class C receptors (reviewed by (Filizola)). A web service, named G-protein coupled Receptors Interaction Partners (GRIP) that predicts the interfaces for 7TMRs oligomerization is also available at http://grip.cbrc.jp/GRIP/index.html (Nemoto et al., 2009). G protein coupled Receptor Interaction Partners DataBase (GRIPDB) has also been developed, which provides information about 7TMRs oligomerization i.e. experimentalaly indentified 7TMRs oligomers, as well as suggested interfaces for the oligomerization (Nemoto et al., 2011).

3.2 Therapeutic application and drug discovery

7TMRs are one of the most important drug targets in the pharmaceutical industry; approximately 40% of the prescription drugs on the market target 7TMRs, but only 5% of the known 7TMR targets are utilized. Agonists and antagonists of 7TMRs are used in the treatment of diseases of every major organ system including the central nervous system, cardiovascular, respiratory, metabolic and urogenital systems. The most exploited 7TMR drug targets include AT_1 angiotensin, adrenergic, dopamine and serotonin (5-hydroxytryptamine, 5-HT) receptor subtypes (Schoneberg et al., 2004). For instance, antagonists of AT_1 angiotensin II receptors are used to prevent diabetes mellitus-induced renal damage and to treat essential hypertension and congestive heart failure. β-adrenergic receptor antagonists, acting on β_1- and/or β_2-adrenergic receptors, are used in patients with congestive heart failure and to treat hypertension and coronary heart disease, while β_2-adrenergic receptor agonists are used in the treatment of asthma, chronic obstructive pulmonary disease and to delay preterm labor. Dopamine receptor antagonists, primarily acting on D_2 receptors, are utilized in the treatment of schizophrenia, while dopamine receptor agonists (e.g. precursor for dopamine levodopa (L-dopa)) remain the standard for treating Parkinson's disease. Inhibitors of 5-HT uptake, which act as indirect agonists at various subtypes of 5-HT receptors, are used to treat major depressive disorders (Schoneberg et al., 2004).

The increasing importance of dimerization for 7TMRs naturally suggests its possible relevance to drug discovery. It seems that the inclination to hetero-dimerize is common among the 7TM members and that the tissue-specific expression patterns probably underlay the creation of relevant receptor pairs. However, 7TMRs expression has been shown to be altered in some pathological situations. In support to the latter preeclampsia was the first disorder linked to alteration in the AT_1–bradykinin B_2 receptor hetero-dimerization (AbdAlla et al., 2001). Opioid and dopamine receptor hetero-dimerization has also been comprehensively studied, since their putative ligands are used in pathological conditions such as basal ganglia disorders, schizophrenia, drug addiction and pain. The increase in the dopamine D_1-D_3 hetero-dimer was shown to be involved in L-dopa-induced dyskinesia in patients with Parkinson's disease and the addition of an adenosine A_{2A} receptor antagonist potentiates the anti-parkinsonian effect of L-dopa. Hetero-dimers of glutamate receptors mGluR2 and 5-HT_{2A} have been specifically associated with hallucinogenic responses in schizophrenia. Furthermore, the opioid δ-μ receptor hetero-dimer is a better target than either μ or δ receptors alone, since blockade of the δ receptor decreases tolerance to the analgesic effects of the most used μ receptor agonist, morphine (reviewed by (Ferré & Franco, 2010; Kenakin & Miller, 2010)). These observations would probably led to broaden the therapeutic potential of drug targeting 7TMRs and it is also anticipated that the evolving concepts of 7TMR dimerization will be implemented in the BRET-based drug discovery and development process (reviewed by (Casado et al, 2009)).

4. BRET principle and its application in the field of 7TMRs dimerization

4.1 BRET principle

BRET is a biophysical method that enables monitoring of physical interactions between two proteins fused to BRET donor and acceptor moieties, respectively, dependent on their intermolecular distance (10 – 100 Å) and on relative orientation due to the dipole-dipole nature of the resonance energy transfer mechanism (Zacharias et al., 2000). BRET is a non-radiative energy transfer, occurring between a bioluminescent donor that emits light in the presence of its corresponding substrate and a complementary fluorescent acceptor, which absorbs light at a given wavelength and re-emits light at longer wavelengths. To fulfill the condition for energy transfer, the emission spectrum of the donor must overlap with the excitation spectrum of the acceptor molecule (Zacharias et al., 2000). BRET occurs naturally in some marine species (e.g. in the sea pansy *Renilla reniformis*) and in 1999, Xu et al. (Xu et al., 1999) utilized this approach to study dimerization of the bacterial Kai B clock protein. Since then, several versions of BRET assays have been developed that use different substrates and/or energy donor/acceptor couples. The original $BRET^1$ technology used the pairing of *Renilla luciferase* (Rluc) as the donor and yellow fluorescent protein (YFP) as the acceptor (Xu et al., 1999; Xu et al., 2003). The addition of coelenterazine h, the natural substrate of *Renilla luciferase* (Rluc), leads to a donor emission of blue light (peak at ~480 nm). When the YFP-tagged acceptor molecule, adapted to this emission wavelength, is in close proximity to the Rluc-tagged donor molecule, excitation of YFP occurs by resonance energy transfer resulting in an acceptor emission of green light (peak at ~530 nm). The substantial overlap in the emission spectra of Rluc and YFP acceptor emission (Stokes shift only ~50 nm) creates a significant problem that has been overcome in a second generation of BRET assay ($BRET^2$). In $BRET^2$ assays, *Renilla luciferase* (Rluc) is used as the donor, the green fluorescent protein (GFP) variant GFP^2 as the acceptor molecule (excitation ~400 nm, emission peak at 510 nm) and the proprietary coelenterazine DeepBlueC™ (also known as coelenterazine 400A) as a substrate. In the presence of DeepBlueC™, Rluc emits light peaking at 395 nm, a wavelength that excites GFP^2 resulting in the emission of green light at 510 nm. This modified BRET pair results in a broader Stokes shift of 115 nm, thus enabling superior separation of donor and acceptor peaks, as well as efficient filtration of the excitation light that it does not come to the detector, thereby enabling detection of the weak fluorescence signal. However, the disadvantage of $BRET^2$, compared to $BRET^1$ is the 100-300 times lower intensity of emitted light and a very fast decay of emitted light (Heding, 2004). $BRET^2$ sensitivity can be improved by the development of suitably sensitive instruments (Heding, 2004) and the use of Rluc mutants with improved quantum efficiency and/or stability (e.g. Rluc8 and Rluc-M) as a donor (De et al., 2007). A third generation BRET assay ($BRET^3$) has been developed recently and combines Rluc8 with the mutant red fluorescent protein (DsRed2) variant mOrange and the coelenterazine or EnduRen™ as a substrate (De et al., 2007). EnduRen™ is a very stable coelenterazine analogue that enables luminescence measurement for at least 24 hours after substrate addition and was utilized in the extended BRET (eBRET) technology (Pfleger et al., 2006). Therefore, in $BRET^3$, donor spectrum is the same as in $BRET^1$, and the red shifted mOrange acceptor signal (emission peak at 564 nm) improves spectral resolution to 85 nm, thereby reducing bleedthrough in the acceptor window. Improved spectral resolution and increased photon intensity allow imaging of protein-protein interactions from intact living cells to small living subjects. Additional optimized donor/acceptor BRET couples that combine Rluc/Rluc8 variant with the yellow fluorescent protein, the YPet variant and the Renilla green fluorescent protein (RGFP) has also been developed (Kamal et al., 2009).

4.2 BRET and 7TMRs dimerization

The use of energy-based techniques such as FRET and BRET has been fundamental for taking the theme of 7TMRs dimerization/oligomerization at the front of 7TMRs research. In 2000, BRET was introduced in the 7TMR field demonstrating β_2-adrenergic receptor (β_2-AR) dimerization (Angers et al., 2000) and since then BRET-based information about 7TMRs homo-/hetero-dimerization is rapidly accumulating (for a recent reviews see (Achour et al., 2011; Ayoub & Pfleger, 2010; Ferré et al., 2009; Ferré & Franco, 2010; Gurevich & Gurevich, 2008a; Gurevich & Gurevich, 2008b; Palczewski, 2010)). As a consequence, knowledge databases have been developed to gather and organize these scattered data and provide researchers with the comprehensive collection of information about 7TMR oligomerization. Existing databases are G protein-coupled receptor oligomer knowledge base (GPCR-OKB) (Skrabanek et al., 2007; Khelashvili et al., 2010) that is freely available at http://www.gpcr-okb.org and G protein-coupled receptor interaction partners database (GRIPDB) (Nemoto et al., 2011) available at http://grip.cbrc.jp/GDB/index.html. By analyzing the data in the GPCR-OKB, we can see that BRET-based approaches were used more often than other experimental approaches such as co-immunoprecipitation, cross-linking, co-expression of fragments or modified protomers, use of dimer specific antibodies, fluorescence resonance energy transfer (FRET) and time resolved FRET to detect oligomerization *in vivo* while in *in vitro* systems others methods still prevail (Table 1). The 7TMR pairs for which functional evidence was provided *in vivo* by BRET are summarized in Table 2. It should be emphasized that besides the intra-family hetero-dimers, the members from different 7TMR families also form functionally relevant inter-family oligomers (Table 2).

Oligomers (*in vivo*)	**7TMR**	**Family A**	**Family B**	**Family C**	**Family A/C**	**Other**
BRET	18	13	0	1	4	0
Mus musculus	7	5	0	1	1	0
Rattus norvegicus	9	5	0	1	3	0
Homo sapiens	9	8	0	0	1	0
Other methods	11	7	0	1	2	1
Oligomers (*in vitro*)						
BRET	50	40	2	1	6	1
Other methods	192	160	4	13	13	2

Table 1. Comparisons of 7TMRs oligomers identified by BRET *vs.* others methods in different 7TMR families in *in vivo* and *in vitro.* Data source GPCR-OKB (http://www.gpcr-okb.org).

Oligomer name	**Organism**	***In vivo* evidence**	**Potential clinical relevance**
Family A 7TMRs			
Adenosine A1 - Adenosine A2A oligomer (**A1 - A2A**)	*Rattus norvegicus*	evidence for physical association in native tissue or primary cells	
Adenosine A2A - Cannabinoid CB1 oligomer (**A2A - CB1**)	*Homo sapiens, Rattus norvegicus*	evidence for physical association in native tissue or primary cells, identification of a specific functional property in native tissue (brain)	Implicated in Parkinson's disease.
Adenosine A2A - Dopamine D2 oligomer (**A2A - D2**)	*Homo sapiens, Rattus norvegicus*	evidence for physical association in native tissue or primary cells, identification of a specific functional property in native tissue (rat striatum, human striatum)	Implicated in Parkinson's desease, schizophrenia. Level of adenosine is increased in the striatal extracellular fluid in Parkinson's disease.
Adrenergic α_1B - Adrenergic α_1D receptor oligomer (**α_1B - α_1D adrenoreceptor**)	*Homo sapiens, Mus musculus*	evidence for physical association in native tissue or primary cells, identification of a specific functional property in native tissue (brain), use of knockout animals or RNAi technology	The study demonstrated that when the α1B-KO and α1D-KO strains of mice are used in conjunction with antagonists, a different pharmacological situation emerges relative to control (sensitivity to Phenylephrine).

Oligomer name	Organism	*In vivo* evidence	Potential clinical relevance
Adrenergic α_2A receptor - Opioid μ receptor oligomer (**α_2A-adrenoreceptor – opioid μ**)	*Homo sapiens*	evidence for physical association in native tissue or primary cells	
Adrenergic β_2 - Prostaglandin EP1 receptor oligomer (**β_2-adrenoreceptor - EP1**)	*Homo sapiens, Mus musculus*	evidence for physical association in native tissue or primary cells, identification of a specific functional property in native tissue (airway smooth muscle)	Implicated in decreasing airway smooth muscle relaxation during asthma.
Cannabinoid CB1 - Dopamine D2 oligomer (**CB1 - D2**)	*Homo sapiens, Rattus norvegicus*	identification of a specific functional property in native tissue	
Chemokine CCR2-CXCR4 receptor oligomer (**CCR2 - CXCR4**)	*Homo sapiens*	identification of a specific functional property in native tissue	
Dopamine D1 - Histamine H3 receptor oligomer (**D1 - H3**)	*Mus musculus*	evidence for physical association in native tissue or primary cells	
Dopamine D1 - Opioid μ receptor oligomer (**D1 – μ**)	*Rattus norvegicus*	evidence for physical association in native tissue or primary cells	
Dopamine D2 - Histamine H3 receptor oligomer (**D2 - H3**)	*Homo sapiens Mus musculus*	evidence for physical association in native tissue or primary cells	
Opioid δ - Opioid κ receptor oligomer (**δ – κ**)	*Mus musculus*	colocalization in spinal cord	tissue-specific agonist for pain
Opioid δ - Opioid μ receptor oligomer (**δ – μ**)	*Mus musculus*	evidence for physical association in native tissue or primary cells, identification of a specific functional property in native tissue	
Family C 7TMRs			
γ-aminobutiric acid GABAb receptor oligomer (**GABAB1 - GABAB2**)	*Rattus norvegicus, Mus musculus*	colocalize in brain	$GABA_{B1}$ agonist Baclofen is an antispasm drug
Family A/C 7TMRs			
Adenosine A2A - Metabotropic glutamate 5 (mGLU 5) oligomer (**A2A - mGLU5**)	*Homo sapiens, Rattus norvegicus*	evidence for physical association in native tissue or primary cells	
Dopamine D2 - Metabotropic glutamate 5 (mGLU 5) oligomer (**D2 - mGLU5**)	*Rattus norvegicus*	evidence for physical association in native tissue or primary cells	

Oligomer name	**Organism**	***In vivo* evidence**	**Potential clinical relevance**
Adenosine A2A - Dopamine D2 - Metabotropic glutamate 5 (mGLU5) oligomer (**A2A - D2 - mGLU5**)	*Rattus norvegicus, Mus musculus*	evidence for physical association in native tissue or primary cells	
Serotonin 5-HT2A receptor oligomer - Metabotropic glutamate 2 (**5-HT2A – mGLU2**)	*Homo sapiens*	evidence for physical association in native tissue or primary cells, identification of a specific functional property in native tissue (brain)	5-HT2A levels increase and mGLU2 levels decrease in schizophrenia

Table 2. Intra- and inter-family oligomers with *in vivo* evidence discovered by BRET method. Data source GPCR-OKB (http://www.gpcr-okb.org).

4.3 Interpretation of BRET results – Possible drawbacks

BRET signal indicates that molecules of the same (or two different) receptors are at maximum distance of 100 Å (that equals 10 nm) or more accurately that the donor and acceptor moieties are within this distance. The efficiency of energy transfer depends on the relative orientation of the donor and acceptor and the distance between them (Zacharias et al., 2000), so that absolute distances can not be measured. Experimentally determined Förster distance R_0 (distance at which the energy transfer efficiency is 50%) for $BRET^1$ and $BRET^2$ is 4.4 nm and 7.5 nm, respectively (Dacres et al., 2010). 7TMR transmembrane core spans ~40 Å across the intracellular surface (Palczewski et al., 2000), which makes BRET suitable to the study of dimerization. However, certain facts need to be considered when interpreting BRET results. Firstly, the size of 27 kDa fluorescent proteins and 34 kDa *Renilla luciferase* is comparable to that of the transmembrane core of 7TMRs (diameter ~40 Å). These proteins are usually attached to the receptor C-terminus, which in different 7TMRs varies in length from 25 to 150 amino acids. Polypeptides of this length in extended conformation can cover 80−480 Å. Thus, a BRET signal indicates that the donor and acceptor moieties are at distance less than 10 nm, which may occur when receptors form structurally defined dimer or when they are far >500 Å apart (reviewed by (Gurevich & Gurevich, 2008a)). The use of acceptor and donor molecules genetically fused to 7TMRs can alter the functionality of the receptor; fusion proteins can also be expressed in the intracellular compartments, thus making difficult to demonstrate that the RET results from a direct interaction of proteins at the cell surface (Ferre & Franco, 2010). The use of fusion proteins can therefore be a major limitation for this application. Secondly, quantitative BRET measurements are limited by the quality of the signal and noise level. Fluorescent proteins and luciferase yield background signals arising from incompletely processed proteins inside the cell and high cell autofluorescence in the spectral region used (Gurevich & Gurevich, 2008a). Thirdly, so called bystander BRET results from frequent encounters between overexpressed receptors and has no physical meaning (Kenworthy & Edidin, 1998; Mercier et al., 2002). BRET assays should therefore be able to discriminate between genuine dimerization compared to random collision due to over-expression. To determine specify of BRET signal the following experiments has been proposed: negative control with a non-interacting receptor or protein, BRET saturation and competition assays and experiments that observe ligand-promoted changes in BRET (Achour et al., 2011; Ayoub

& Pfleger, 2010; Ferre & Franco, 2010). Additionally, interpretation of BRET data also requires quantitative analysis of the results, which was so far done only in a small number of studies (Ayoub et al., 2002; Mercier et al., 2002; Vrecl et al., 2006). The theoretical background of the assays described below provides some guidelines for the appropriate interpretation and quantitative evolution of BRET results.

5. Mathematical models to quantitatively assess the oligomerization state of studied receptors

5.1 Basic assumptions

Bioluminescent resonance energy transfer takes place at 1-10 nm distances between molecules thus allowing study of protein-protein interaction. It is a quite robust tool but still some care should be taken with interpretation of the results. Resonance energy transfer is described by the Förster equation for energy transfer efficiency E (Förster, 1959):

$$E = \frac{R_0^6}{R_0^6 + r^6} \tag{1}$$

where r is a distance between donor and acceptor, Förster radius R_0 depends on spectral overlap and dipole orientations yielding R_0 values of 4.4 nm for $BRET^1$and 7.5 nm for $BRET^2$ (Dacres et al., 2010). E is an important parameter in interpretation of the BRET assays used for oligomerisation studies. If the BRET luminometer is properly calibrated then E can be calculated from the $BRET_{max}$ signal obtained when all donor molecules are accompanied by acceptor molecules:

$$E = \frac{BRET_{max}}{BRET_{max} + 1} \tag{2}$$

Calibration should take into account differences in the detector quantum efficiencies at donor and acceptor emission wavelengths and the proportion of the detected emission spectra of both markers. Knowing a Förster radius for certain type of BRET technology used and energy transfer efficiency E we can estimate the distance between the donor and acceptor marker species in the protein complex.

Calculations in presented BRET assays are derived from Veatch and Stryer article (Veatch & Stryer, 1977) covering FRET experiments with Gramicidin dimers. In FRET experiments the 28 $Q/Q0$ is a measurement parameter representing the ratio between not-transmitted energy Q and total energy Q_0. Vaecht and Stryer equations have been adopted for BRET experiments where we measure the ratio between transmitted T and not-transmitted energy Q:

$$BRET = \frac{T}{Q} = \frac{Q_0}{Q} - 1 \tag{3}$$

Single BRET measurements do not give unambiguous proof that receptors form oligomers because the signal can be a consequence of random collisions. To get better indication of the oligomerisation state several quantitative assays were developed.

5.2 BRET dilution assay

This is a simplest control experiment to check for oligomerisation. Resonant energy transfer takes place if the distance between donor and acceptor molecules is in the range of Förster radius R_0. Molecules can get close enough for BRET also by random collisions (bystander BRET) if their density is high enough (Kenworthy & Edidin, 1998; Mercier et al., 2002). Excluding random collisions there should be no concentration dependence for coupled donor and acceptor molecules. In practice we can approximate the *BRET* signal as:

$$BRET = BRET_0 + k([D]+[A]) \quad (4)$$

where $[D]$ and $[A]$ are donor and acceptor concentrations. With lowering the concentration of both receptors simultaneously (dilution) the *BRET* signal approaches $BRET_0$ which is the real oligomerisation signal (Fig. 1). Dilution assay is used to set the concentration range for saturation and competition assays (Breit et al., 2004).

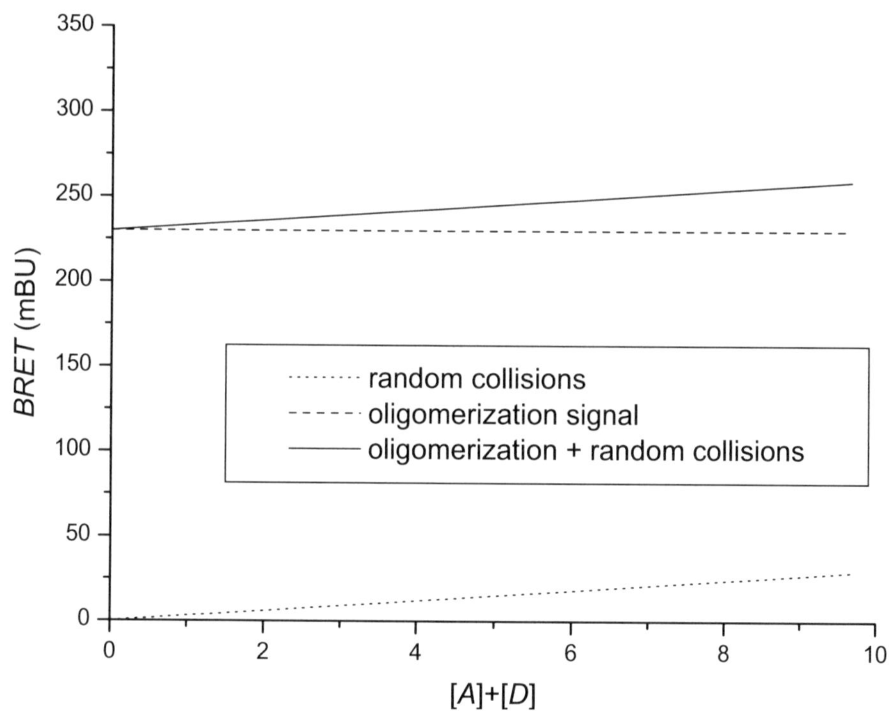

Fig. 1. BRET dilution assay. Theoretical BRET concentration curves for receptors forming monomers or oligomers. A constant ratio between acceptor and donor concentrations should be used.

5.3 BRET saturation assay

Saturation assay involves expressing a constant amount of donor-tagged receptor with an increasing amounts of acceptor-tagged receptor. Theoretically, *BRET* signal should increase with increasing amounts of acceptor until all donor molecules are interacting with acceptor molecules. Therefore, a saturation level is achieved beyond which a further elevation of the amount of acceptor does not increase the *BRET* signal, thereby reaching a maximal *BRET* level ($BRET_{max}$) (Achour et al., 2011; Ayoub & Pfleger, 2010; Hamdan et al., 2006; Mercier et al., 2002). By using a saturation assay it is possible to obtain the oligomerisation state of homologous receptors. BRET saturation curve is derived from Veatch and Stryer model:

$$BRET = \frac{T}{Q} = \frac{E[AD]}{2[DD] + (1-E)[AD]} \tag{5}$$

where $[AD]$ are acceptor-donor and $[DD]$ donor-donor dimer concentrations. If all receptors form dimers and association constants are the same for AA, AD and DD we obtain BRET saturation curve for dimers:

$$BRET = \frac{E\frac{[A]}{[D]}}{1 + (1-E)\frac{[A]}{[D]}} \tag{6}$$

For higher oligomers a general BRET saturation curve can be derived (Vrecl et al., 2006):

$$\frac{BRET}{BRET_{max}} = 1 - \frac{1}{E + (1-E)\left(1 + \frac{[A]}{[D]}\right)^{N}} \tag{7}$$

where N=1 for dimer, N=2 for trimer and N=3 for tetramer. Theoretical BRET saturation curves are presented in Fig. 2. *BRET* for higher oligomers shows faster saturation. For comparison the monomer *BRET* signal which corresponds to random collisions is presented. If receptor concentration is very high then random collisions can generate saturation curve similar to that of the dimers. Thus a dilution experiment should be done first to distinguish random collisions from the oligomerisation.

In heterologous saturation assay different receptors are used as donors and acceptors. In this case saturation curve is influenced by the affinities for homo-dimer and hetero-dimer formation. In practice we can observe a right-shift of the saturation curve where the association constant for hetero-dimers is smaller than that of the homo-dimers yielding higher $BRET_{50}$ values.

5.4 BRET competition assay

In an attempt to further confirm the existence of oligomer complexes, competition assay can be performed. In this assay the concentration of untagged receptor is increased over a constant concentrations of donor and acceptor tagged receptors (Achour et al., 2011; Vrecl et al., 2006). It is expected that the *BRET* signal would decrease if untagged receptors compete with the tagged receptors for the binding in complexes. Following the Veatch and Stryer approach we obtain *BRET* signal:

$$BRET = \frac{T}{Q} = \frac{E[AD]}{2[DD] + (1-E)[AD] + [CD]} \tag{8}$$

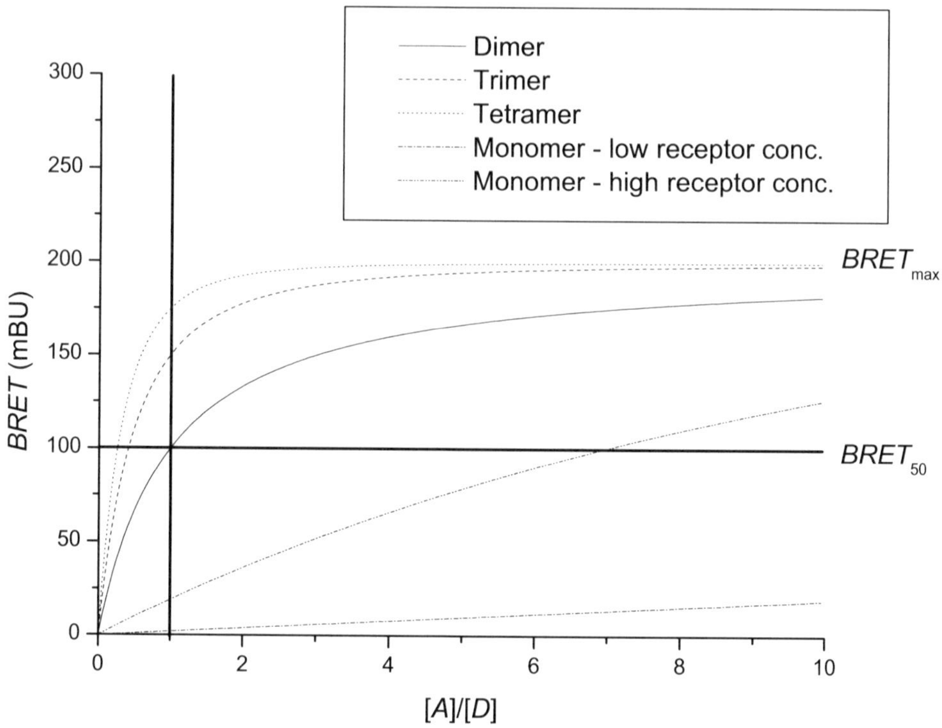

Fig. 2. BRET saturation assay. Theoretical curves for oligomer formation are plotted as a function of ratio of receptors tagged with acceptor [*A*] and donor [*D*] molecules. In the case of monomers the BRET signal is created due to random collisions.

where C represents untagged competitor. If all receptors form dimers and association constants are the same for AA, AD, DD, CD, AC and CC dimers we obtain BRET competition curve for dimers:

$$BRET = \frac{E\frac{[A]}{[D]}}{1+(1-E)\frac{[A]}{[D]}+\frac{[C]}{[D]}} \tag{9}$$

Usually in BRET saturation experiments high acceptor to donor concentration ratio is used because the variation in this ratio do not influence the *BRET* signal as much as for $[A]/[D]$=1. In general the interaction with the untagged receptors causes the reduction of *BRET* signal following a hyperbolic curve (Figure 3). We can very well distinguish if the oligomerisation is present, but the exact oligomerisation state is difficult to assess. Competition assay is more suited for the study of hetero-oligomers where different kind of untagged receptor is competing with the homo-oligomers. The saturation curve is shallower if there is a low affinity for hetero-dimer formation compared to homo-dimers

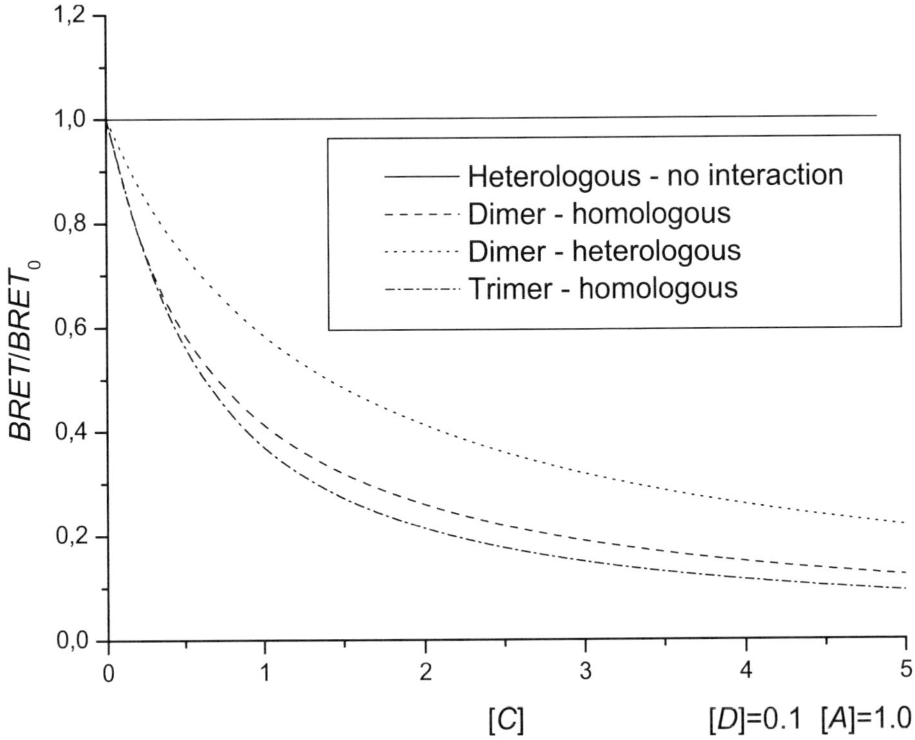

Fig. 3. BRET competition assay. In homologous assay the same receptor is used as a competitor, whereas in heterologous assay different receptor is used. For the latter case a hetero-dimer with lower association constant than that of the homo-dimer is presented.

6. Other BRET-based approaches to identify 7TMR hetero-dimerization

To overcome certain limitations of the classical BRET assays described above, some other BRET-based approaches have been developed to study 7TMR oligomerization/ hetero-dimerization. Sequential-BRET-FRET (SRET) enables identification of oligomers formed by three different proteins. In SRET, the oxidation of the RLuc substrate by an RLuc-fusion protein triggers the excitation of the acceptor GFP^2 by $BRET^2$ and subsequent energy transfer to the acceptor YFP by FRET. Combination of bimolecular fluorescence complementation (BiFC) and BRET techniques is based on the ability to produce a fluorescent complex from non-fluorescent constituents if a protein-protein interaction occurs. Two receptors are fused at their C-termini with either N-terminal or C-terminal fragments of YFP, respectively, and receptor hetero-dimerization causes YFP reconstitution. Then, if there is hetero-trimerization, BRET can be obtained when the cells also co-express the third receptor fused to Rluc (reviewed by (Ferré & Franco, 2010)). GPCR-Heteromer Identification Technology (GPCR-HIT) utilizes BRET and ligand-dependent recruitment of a 7TMR-specific interaction partners (such as a β-arrestin, PKC or G-protein) to enable 7TMR heteromer discovery and characterization (Mustafa & Pfleger, 2011; See et al., 2011). In this set up, only one receptor subtype is fused to Rluc and the second receptor subtype is untagged. A third protein capable of interacting specifically with one or both receptors in a ligand-dependent manner is fused to a YFP. Ligand-induced BRET signal indicates that activation of untagged receptor or the heteromer results in recruitment of YFP-tagged protein to the heteromer. Recently developed complemented donor-acceptor resonance energy transfer (CODA-RET) method combines protein complementation with resonance energy transfer to study conformational changes in response to activation of a defined G protein-coupled receptor heteromer. CODA-RET quantify the BRET between a receptor hetero-dimer and a subunit of the heterotrimeric G-protein. It eliminates a contribution from homodimeric signaling and enables analyzing the effect of drugs on a defined 7TMR heterodimer (Urizar et al., 2011).

7. Conclusions

BRET-based techniques are extremely powerful, provided that they are conducted with the appropriate controls and correctly interpreted. Quantitative BRET assays allow us to support the ability of receptor for homo-dimer and hetero-dimer. Homologous saturation assay provide us with the oligomerisation state of receptors. Data interpretation is more difficult for hetero-oligomers and the mixtures of monomer, dimer and higher oligomer populations. For the quantitative approach we also need to know the relative concentrations of all receptors used in the experiment, which can be obtained from radioligand binding, Western blot or ELISA assays.

8. Acknowledgment

We acknowledge funding from the Slovenian Research Agency (program P4-0053) and Slovenian-Danish collaboration grants (BI-DK/06-07-007, BI-DK/07-09-002 and BI-DK/11-12-008).

9. References

AbdAlla S, Lother H, el Massiery A, Quitterer U. (2001) Increased AT(1) receptor heterodimers in preeclampsia mediate enhanced angiotensin II responsiveness. *Nat Med* 7, 1003-1009.

Achour L, KM, Jockers R, Marullo S. (2011) Using quantitative BRET to assess G protein-coupled receptor homo- and heterodimerization. *Methods Mol Biol*, 756: 183-200.

Angers S, Salahpour A, Joly E et al. (2000) Detection of beta 2-adrenergic receptor dimerization in living cells using bioluminescence resonance energy transfer (BRET). *Proc Natl Acad Sci U S A*, 97: 3684-3689.

Avissar S, Amitai G, Sokolovsky M. (1983) Oligomeric structure of muscarinic receptors is shown by photoaffinity labeling: subunit assembly may explain high- and low-affinity agonist states. *Proc Natl Acad Sci U S A*, 80: 156-159.

Ayoub MA, Couturier C, Lucas-Meunier E et al. (2002) Monitoring of ligand-independent dimerization and ligand-induced conformational changes of melatonin receptors in living cells by bioluminescence resonance energy transfer. *J Biol Chem*, 277: 21522-21528.

Ayoub MA, Pfleger KD. (2010) Recent advances in bioluminescence resonance energy transfer technologies to study GPCR heteromerization. *Curr Opin Pharmacol*, 10: 44-52.

Breit A, Lagace M, Bouvier M. (2004) Hetero-oligomerization between beta2- and beta3-adrenergic receptors generates a beta-adrenergic signaling unit with distinct functional properties. *J Biol Chem*, 279: 28756-28765.

Casadó V, Cortés A, Mallol J, Pérez-Capote K, Ferré S, Lluis C, Franco R, Canela EI. (2009) GPCR homomers and heteromers: a better choice as targets for drug development than GPCR monomers? *Pharmacol Ther*, 124: 248-257.

Dacres H, Wang J, Dumancic MM, Trowell SC. (2010) Experimental determination of the Forster distance for two commonly used bioluminescent resonance energy transfer pairs. *Anal Chem*, 82: 432-435.

De A, Loening AM, Gambhir SS. (2007) An improved bioluminescence resonance energy transfer strategy for imaging intracellular events in single cells and living subjects. *Cancer Res*, 67: 7175-7183.

De Meyts P. (1976) Cooperative properties of hormone receptors in cell membranes. *J Supramol Struct*, 4: 241-258.

de Meyts P, Roth J, Neville DM, Jr., Gavin JR, 3rd, Lesniak MA. (1973) Insulin interactions with its receptors: experimental evidence for negative cooperativity. *Biochem Biophys Res Commun*, 55: 154-161.

Ferré S, Baler R, Bouvier M et al. (2009) Building a new conceptual framework for receptor heteromers. *Nat Chem Biol*, 5: 131-134.

Ferré S, Franco R. (2010) Oligomerization of G-protein-coupled receptors: a reality. *Curr Opin Pharmacol*, 10: 1-5.

Filizola M. (2010) Increasingly accurate dynamic molecular models of G-protein coupled receptor oligomers: Panacea or Pandora's box for novel drug discovery? *Life Sci* 2010, 86: 590-597.

Förster T. (1959) Transfer mechanisms of electronic excitation. *Discuss. Faraday Soc.* 27: 7-17.

Fuxe K, Agnati LF, Benfenati F et al. (1983) Evidence for the existence of receptor--receptor interactions in the central nervous system. Studies on the regulation of monoamine receptors by neuropeptides. *J Neural Transm Suppl*, 18: 165-179.

Gonzalez-Maeso J, Ang RL, Yuen T et al. (2008) Identification of a serotonin/glutamate receptor complex implicated in psychosis. *Nature*, 452: 93-97.

Gurevich VV, Gurevich EV. (2008a) GPCR monomers and oligomers: it takes all kinds. *Trends Neurosci*, 31: 74-81.

Gurevich VV, Gurevich EV. (2008b) How and why do GPCRs dimerize? *Trends Pharmacol Sci*, 29: 234-240.

Gurevich VV, Gurevich EV. (2006) The structural basis of arrestin-mediated regulation of G-protein-coupled receptors. *Pharmacol Ther*, 110: 465-502.

Hamdan FF, Percherancier Y, Breton B, Bouvier M. (2006) Monitoring protein-protein interactions in living cells by bioluminescence resonance energy transfer (BRET). *Curr Protoc Neurosci*, Chapter 5: Unit 5 23.

Heding A. (2004) Use of the BRET 7TM receptor/beta-arrestin assay in drug discovery and screening. *Expert Rev Mol Diagn*, 4: 403-411.

Hucho F, Tsetlin V. (1996) Structural biology of key nervous system proteins. *J Neurochem*, 66: 1781-1792.

Kamal M, Marquez M, Vauthier V et al. (2009) Improved donor/acceptor BRET couples for monitoring beta-arrestin recruitment to G protein-coupled receptors. *Biotechnol J*, 4: 1337-1344.

Kenakin T, Miller LJ. (2010) Seven transmembrane receptors as shapeshifting proteins: the impact of allosteric modulation and functional selectivity on new drug discovery. *Pharmacol Rev*, 62: 265-304.

Kenworthy AK, Edidin M. (1998) Distribution of a glycosylphosphatidylinositol-anchored protein at the apical surface of MDCK cells examined at a resolution of <100 Å using imaging fluorescence resonance energy transfer. *J Cell Biol*, 142: 69-84.

Khelashvili G, Dorff K, Shan J et al. (2010) GPCR-OKB: the G Protein Coupled Receptor Oligomer Knowledge Base. *Bioinformatics*, 26: 1804-1805.

Kniazeff J, Prezeau L, Rondard P, Pin JP, Goudet C. (2011) Dimers and beyond: The functional puzzles of class C GPCRs. *Pharmacol Ther*, 130: 9-25.

Kyte J, Doolittle RF. (1982) A simple method for displaying the hydropathic character of a protein. *J Mol Biol*, 157: 105-132.

Limbird LE, Meyts PD, Lefkowitz RJ. (1975) Beta-adrenergic receptors: evidence for negative cooperativity. *Biochem Biophys Res Commun*, 64: 1160-1168.

McGraw DW, Liggett SB: G-protein coupled receptors. (2006) In Laurent GJ, Shapiro SD (eds.) Encyclopedia of Respiratory Medicine. Oxford: Elsevier Ltd.; 248-251.

Mercier JF, Salahpour A, Angers S, Breit A, Bouvier M. (2002) Quantitative assessment of beta 1- and beta 2-adrenergic receptor homo- and heterodimerization by bioluminescence resonance energy transfer. *J Biol Chem*, 277: 44925-44931.

Mustafa S, Pfleger KD. (2011) G protein-coupled receptor heteromer identification technology: identification and profiling of GPCR heteromers. *J Lab Autom*, 16: 285-291.

Nemoto W, Fukui K, Toh H. (2009) GRIP: a server for predicting interfaces for GPCR oligomerization. *J Recept Signal Transduct Res*, 29: 312-317.

Nemoto W, Fukui K, Toh H. GRIPDB - G protein coupled Receptor Interaction Partners DataBase. *J Recept Signal Transduct Res*, 31: 199-205.

Palczewski K. (2010) Oligomeric forms of G protein-coupled receptors (GPCRs). *Trends Biochem Sci*, 35: 595-600.

Palczewski K, Kumasaka T, Hori T et al. (2000) Crystal structure of rhodopsin: A G protein-coupled receptor. *Science*, 289: 739-745.

Pfleger KD, Dromey JR, Dalrymple MB, Lim EM, Thomas WG, Eidne KA. (2006) Extended bioluminescence resonance energy transfer (eBRET) for monitoring prolonged protein-protein interactions in live cells. *Cell Signal*, 18: 1664-1670.

Rivero-Muller A, Chou YY, Ji I et al. (2010) Rescue of defective G protein-coupled receptor function in vivo by intermolecular cooperation. *Proc Natl Acad Sci U S A*, 107: 2319-2324.

Salon JA, Lodowski DT, Palczewski K. (2011) The significance of g protein-coupled receptor crystallography for drug discovery. *Pharmacol Rev*, 63: 901-937.

Schoneberg T, Schulz A, Biebermann H et al (2004) Mutant G-protein-coupled receptors as a cause of human diseases. *Pharmacol Ther* 104: 173-206.

See HB, Seeber RM, Kocan M, Eidne KA, Pfleger KD. (2011) Application of G protein-coupled receptor-heteromer identification technology to monitor beta-arrestin recruitment to G protein-coupled receptor heteromers. *Assay Drug Dev Technol*, 9: 21-30.

Skrabanek L, Murcia M, Bouvier M et al. (2007) Requirements and ontology for a G protein-coupled receptor oligomerization knowledge base. *BMC Bioinformatics*, 8: 177.

Terrillon S, Bouvier M. (2004) Roles of G-protein-coupled receptor dimerization. *EMBO Rep*, 5: 30-34.

Urizar E, Yano H, Kolster R, Gales C, Lambert N, Javitch JA. (2000) CODA-RET reveals functional selectivity as a result of GPCR heteromerization. *Nat Chem Biol*, 7: 624-630.

Veatch W, Stryer L. (1977) The dimeric nature of the gramicidin A transmembrane channel: conductance and fluorescence energy transfer studies of hybrid channels. *J Mol Biol*, 113: 89-102.

Vrecl M, Drinovec L, Elling C, Heding A. (2006) Opsin oligomerization in a heterologous cell system. *Journal of Receptors and Signal Transduction*, 26: 505-526.

White JH, Wise A, Main MJ et al. (1998) Heterodimerization is required for the formation of a functional GABA(B) receptor. *Nature*, 396: 679-682.

Xu Y, Kanauchi A, von Arnim AG, Piston DW, Johnson CH. (2003) Bioluminescence resonance energy transfer: monitoring protein-protein interactions in living cells. *Methods Enzymol*, 360: 289-301.

Xu Y, Piston DW, Johnson CH. (1999) A bioluminescence resonance energy transfer (BRET) system: application to interacting circadian clock proteins. *Proc Natl Acad Sci U S A*, 96: 151-156.

Zacharias DA, Baird GS, Tsien RY. (2000) Recent advances in technology for measuring and manipulating cell signals. *Curr Opin Neurobiol*, 10: 416-421.

5

Use of ATP Bioluminescence for Rapid Detection and Enumeration of Contaminants: The Milliflex Rapid Microbiology Detection and Enumeration System

Renaud Chollet and Sébastien Ribault
Merck-Millipore
France

1. Introduction

Rapid microbial detection becomes increasingly essential to many companies in pharmaceutical, clinical and in food and beverage areas. Faster microbiological methods are required to contribute to a better control of raw materials as well as finished products. Rapid microbiological methods can also provide a better reactivity throughout the manufacturing process. Implementing rapid technologies would allow companies for cost saving and would speed up products release. Despite clear advantages, traditional methods are still widely used. Current methods require incubation of products in liquid or solid culture media for routinely 2 to 7 days before getting the contamination result. This necessary long incubation time is mainly due to the fact that stressed microorganisms found in complex matrices require several days to grow to visible colonies to be detected. Moreover, this incubation period can be increased up to 14 days in specific application like sterility testing for the release of pharmaceutical compounds. Although these techniques show advantages like simplicity, the use of inexpensive materials and their acceptability to the regulatory authorities, the major drawback is the length of time taken to get microbiological results. Thus, face to the growing demand for rapid detection methods, various alternative technologies have been developed. In the field of rapid microorganisms detection, ATP-bioluminescence based on luciferine/luciferase reaction has shown great interest. Indeed, adenosine triphosphate (ATP) is found in all living organisms and is an excellent marker for viability and cellular contamination. Detection of ATP through ATP-luminescence technology is therefore a method of choice to replace traditional method and significantly shorten time to detection without loosing reliability.

This chapter will address the ATP-bioluminescence principle as a sensitive and rapid detection technology in the Milliflex® Rapid Microbiology Detection and Enumeration System (RMDS). This system combines membrane filtration principle, detection of microorganisms by ATP-bioluminescence and light capture triggered by a Charged Coupled Device camera (CCD) followed by software analysis.

2. ATP-Bioluminescence

2.1 ATP-Bioluminescence principle

Light-producing living organisms are widespread in nature and from diverse origins. The process of light emission from organisms is called bioluminescence and represents a chemical conversion of energy into light. Since the work of William D McElroy showing that ATP is a limiting and key factor of the bioluminescent reaction, research has lead to a better understanding of how light is produced by fireflies (McElroy, 1947; McElroy, 1951; McElroy et al., 1953). The bioluminescence mechanism involving Luciferase enzyme is a multistep process which mainly requires Luciferin substrat, Oxygen (O2), Magnesium cation (Mg++) and ATP (DeLuca & McElroy, 1974; McElroy et al., 1953; Seliger, 1989). ATP-bioluminescence using luciferine/luciferase relies on luciferine oxidation by the luciferase and the integrated light intensity is directly proportional to ATP contents. Luciferase converts in presence of ATP and Magnesium firefly D-luciferin into the corresponding enzyme-bound luciferil adenylate. The luciferil adenylate complex is then the substrate of the subsequent oxidative reaction leading to oxyluciferin. The light emission is a consequence of a rapid loss of energy of the oxyluciferine molecule from an excited state to a stable one. This reaction induces the emission of photons with a efficient quantum yield of about 90% (Seliger, 1989; Wilson & Hasting, 1998) (Fig1).

1/ D-luciferin + luciferase + ATP $\xrightarrow{Mg^{++}}$ Luciferil adenylate complex +PPi

2/ Luciferil adenylate complex $\xrightarrow{O_2}$ Oxyluciferin + AMP+ CO2 + light

Fig. 1. Chemical reactions of the ATP-bioluminescence based on luciferin/luciferase system (PPi:inorganic pyrophosphate, CO2: Carbon Dioxide). Photons of yellow-green light (550 to 570 nm) are emitted.

2.2 Luciferase protein

Luciferase is a common term used to describe enzymes able to catalyze light emission. Luciferase belongs to the adelynate-forming protein family and is an oxygen-4-oxidoreductase gathering decarboxylation and ATP-hydrolysing main activities. Structural studies have shown that Photinus pyralis Luciferase protein is folded into 2 domains: a large N-terminal body and a small C-terminal domain linked by a flexible peptide creating a wide cleft (Conti et al., 1996). Amino acids critical for bioluminescence phenomenon belong mainly to the N-terminal domain (Branchini et al., 2000; Thompson et al., 1997; Zako et al., 2003). This implies that luciferine-binding site is mediated by conformational change to bring the 2 domains closer. This conformational change is consistent with the study of Nakatsu et al (2006) showing that luciferase from luciola cruciata exists in an "open form" and in a "closed form", the later form creates an hydrophobic pocket around the active site and is responsible of light emission. Two kinds of colored light emission are described for luciferine/luciferase reaction. The typical high energy yellow-green light emission with a peak at 562 nm at pH 7.5 and red light emission with a peak at 620nm when the pH decreases to 5 (Seliger et al., 1964; Seliger & McElroy, 1964). This surprising phenomenon where Luciferase is able to emit light of different colors is not clearly understood but the isolation of colored luciferase variants shows that single amino acid substitution in

N-terminal domain affects bioluminescence color by modulating slightly the polarity of the active site environment (Hosseinkhani, 2011; Shapiro et al., 2005). This interesting feature opens the way to wide applications in biotechnology (Branchini et al., 2005).

2.3 ATP-Bioluminescence applications

With the isolation, cloning and purification of various luciferases from many bioluminescence-producing organisms (bacteria, beetles, marines organisms, etc), bioluminescent assays have been developed and widely used in microbiology to detect bacterial contamination by measuring presence of ATP and in molecular and cellular biology with luciferase as reporter gene to monitor gene expression, protein-protein interaction, etc (Francis et al., 2000; Roda et al, 2004; Thorne et al., 2010). The average intracellular ATP content in various microorganisms has been quantified and ATP has been shown to be a reliable biomarker of the presence of living organisms (Kodata et al., 1996; Thore et al., 1975; Venkateswaran et al., 2003). To be able to specifically detect living organisms by ATP-bioluminescence, the first step is to extract ATP from cells. This step is critical and impacts directly the reliability of the detection (Selan et al., 1992). Chemical solution or physical extraction methods were used in liquid samples (Selan et al., 1992; Siro et al., 1982). Some false negative results were described in few studies (Conn et al., 1975; Kolbeck et al., 1985). Additional studies investigated the cause of false negative results and demonstrated that ATP extraction was not efficient. Indeed, extensive sonication of bacterial samples for instance caused a significant increase of Relative Light Unit (RLU) measured (Selan et al., 1992). Taking into account this limitation, ATP-bioluminescent assay has already proved to provide good detection properties in many areas. Bioluminescent assay is broadly used to monitor air and surface cleanliness and product quality mainly in food industries and in less extent in pharmaceutical industries (Aycicek et al., 2006; Bautisda et al., 1995; Davidson et al., 1999; Dostalek & Branyik, 2005; Girotti et al., 1997; Hawronskyj & Holah, 1999). Studies shows that the level of contamination assessed though surface swabbing, ATP extraction and bioluminescent assay correlate well for 80 % of the samples tested with traditional plate method (Poulis et al., 1993). Availability of sensitive luminometers as well as many commercial ATP-bioluminescent kits has allowed the development of various protocols and applications in industrial microbiology. Currently, ATP- bioluminescence is an accepted and common technology used to monitor contamination in areas such as food and beverage, ecology, cosmetic, and clinical (Andreotti & Berthold, 1999; Chen & Godwin, 2006; Davidson et al., 1999; Deininger & Lee, 2001; Frundzhyan & Ugarova, 2007; Miller et al., 1992; Nielsen & Van Dellen, 1989; Selan et al., 1992; Yan et al., 2011).

3. Milliflex rapid microbiological detection and enumeration system

3.1 System description

RMDS offers a way to detect and quantify living microorganisms grown on a membrane. By combining ATP-bioluminescence and sensitive detection system, the microbial detection is obtained more rapidly than traditional method. In order to detect a colony or a micro-colony on a membrane by ATP-bioluminescence, the first step is to release ATP from cells. This critical step is achieved by nebulizing automatically an ATP-releasing solution onto the membrane. ATP extraction is made on microcolonies grown on membrane which represents an advantage compared to chemical or physical extraction in liquid. Once ATP is released from lysed cells, it becomes accessible to bioluminescent reaction. A second solution is then automatically nebulized onto the same membrane. This solution brings to lysed cells all components, except ATP, involved in the Luciferin/Luciferase bioluminescence chemical reaction. A spray station is used to uniformly apply small volumes of reagents onto the membrane. As soon as bioluminescent reagents are sprayed onto the membrane, the bioluminescence reaction starts and photons are emitted. The membrane is then transferred manually from the spray station to the detection system. The Milliflex Rapid detection system combines the use of a highly sensitive CCD camera to monitor light emitted from microorganisms and an image analysis software to analyze the signal and give the number of microorganisms counted. The figure 2 shows the detection tower components and their function.

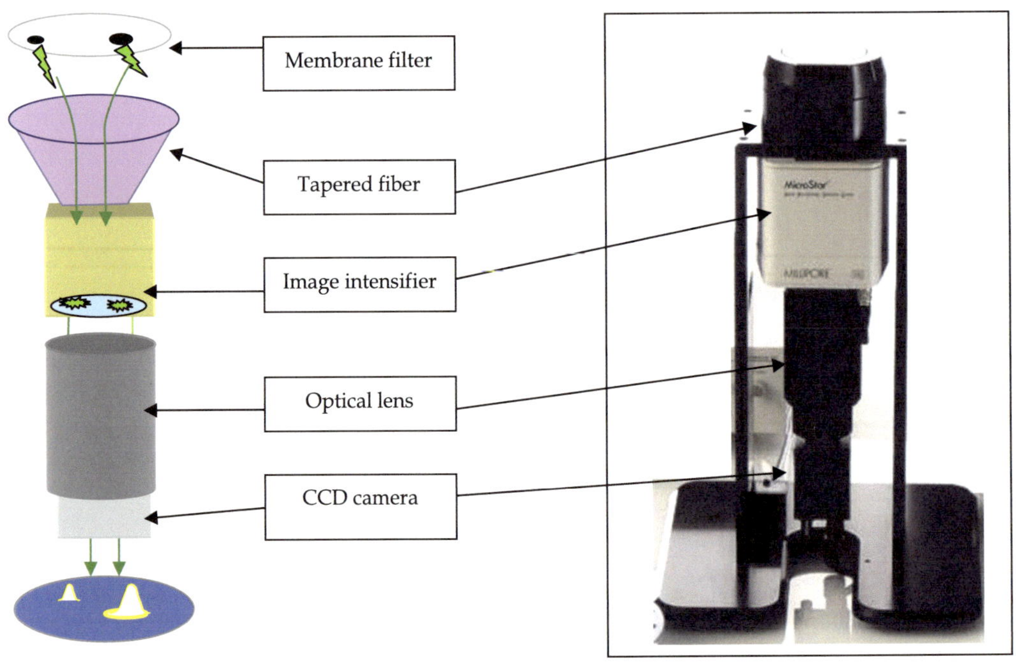

Fig. 2. Milliflex detection tower components: RMDS collects, amplifies, and registers on a CCD camera the light activity of bioluminescent reaction. Photons emitted by microorganisms go through the tapered fiber in order the light to be concentrated and becomes compatible with the size diameter of the CCD camera. In the intensifier, photons hit a photocathode and each photon is converted into cloud of electrons. Then electrons hit a phosphorous screen and are converted back into photons. The CCD camera records light every 30 times per second.

Data collected by the CCD camera are analyzed and treated by software to build an image of the membrane loaded on the top of the detection tower. The image indicates the place where light is emitted. As the signal is collected over a short period (integration time), spots size on the picture represents the light intensity accumulated or emitted by microorganisms (Fig.3).

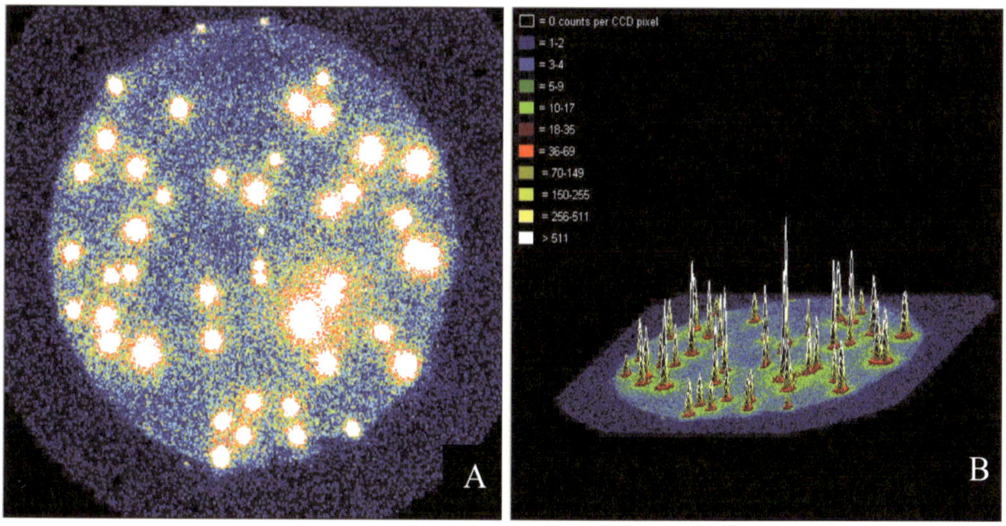

Fig. 3. Example of image given by RMDS software. Picture show the image of the membrane with spots (A) or peaks in 3 dimensions (B) representing exactly the place on the membrane where light is emitted. The result in colony forming unit is directly given by the system.

3.2 RMDS ATP-Bioluminescence protocol

The RMDS ATP-bioluminescence protocol includes the following steps:

1. filter the sample through a Milliflex funnel; 2. incubate the sample onto media; 3. separate the membrane from the media and let the membrane dry inside a laminar flow hood; 4. spray the ATP-releasing reagent and bioluminescence reagent onto the membrane by means of the Milliflex Rapid Autospray Station. The reaction between the ATP from microorganisms and the luciferase enzyme produces light; 5. place the membrane onto the detection tower and initiate detection and enumeration. Photons are detected by the system via a photon counting imaging tube coupled to a CCD camera. The photons generated by the ATP bioluminescence reaction are captured, and the integrated picture is displayed on the computer monitor; 6. after data treatment, a picture of the membrane is provided in two dimensions (2-D) exhibiting spots that represent colonies and in three dimensions (3-D) with peaks that correlate with the ATP content of the colony. The result is directly displayed in colony-forming unit (cfus)on the software screen. The successive steps are summarized in Fig. 4.

The standard protocol, performed in parallel, includes the following steps:

1. filter the sample through a Milliflex funnel; 2. incubate the sample and visually count cfus after incubation.

3.3 Evaluation of Luciferin/Luciferase relative concentrations for optimal detection of microorganisms

The relative concentrations of the 2 key components of the detection reagents were evaluated.

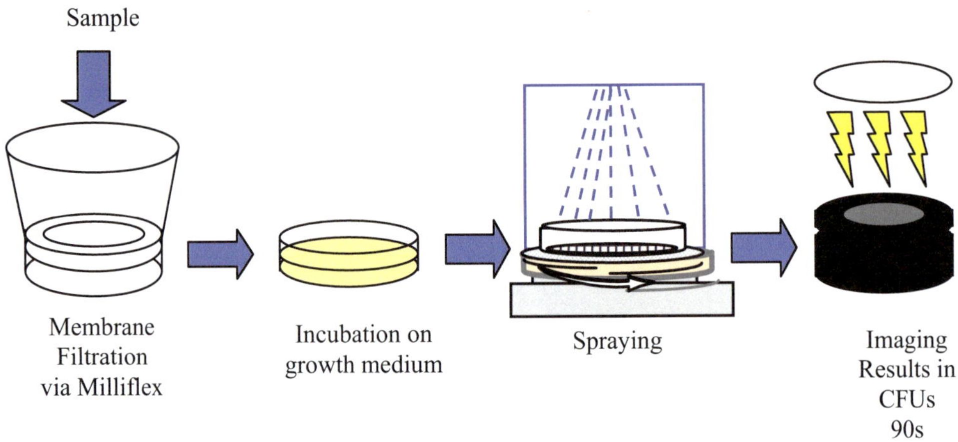

Fig. 4. RMDS ATP-bioluminescence protocol

The protocol used is described in the previous paragraph "RMDS ATP bioluminescence protocol". Only the reagent used for detection varies for the 2 components relative concentrations as described in table 1.

	Formulation 1	Formulation 2	Formulation 3	Formulation 4	Formulation 5
Luciferase	3x	1.5x	1x	1.5x	1x
Luciferin	1x	1x	1x	0.5x	0.5x

Table 1. Formulations relative concentrations of Luciferin/Luciferase tested

The signal and background were determined using membranes incubated during 6h at 32.5°C on Tryptic Soy Agar inoculated with *Escherichia coli* or *Staphylococcus aureus* (table 2).

Formulation 1 gave a signal so strong that the detection system was almost saturated. This saturation did not allow the accurate detection of bacteria on the membrane. The same issue occurred to a weaker extent using formulation 2. On the other hand, while the detection of *S. aureus* was accurate using formulation 5, the signal was too weak to allow all colonies of *E. coli* to be counted. Formulations 3 and 4 were both able to generate a good signal associated with low background. We can conclude from these results that the luciferin and luciferase concentration can be increased to optimize the signal but also that the balance between the 2 components is key. Signal will be increased while increasing concentrations but background as well. Formulation 3 which benefits from the best signal on background ratio has been used during the rest of the studies presented here. It is noticeable that depending on the application, the type of sample tested and the resulting background, this luciferase to luciferin balance can be adjusted to better match the detection criteria and increase signal on background ratio.

	E. coli	*S. aureus*
Formulation 1	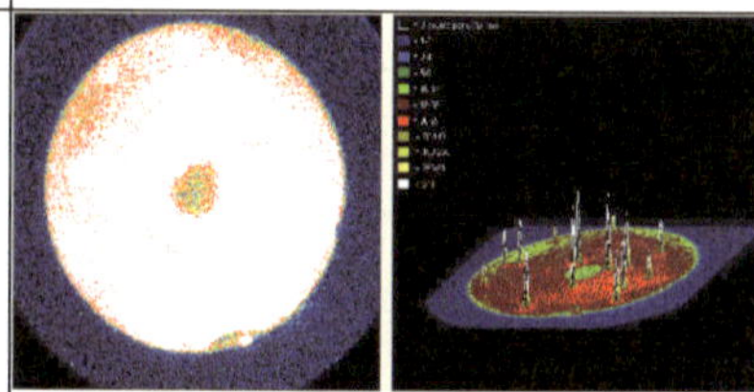	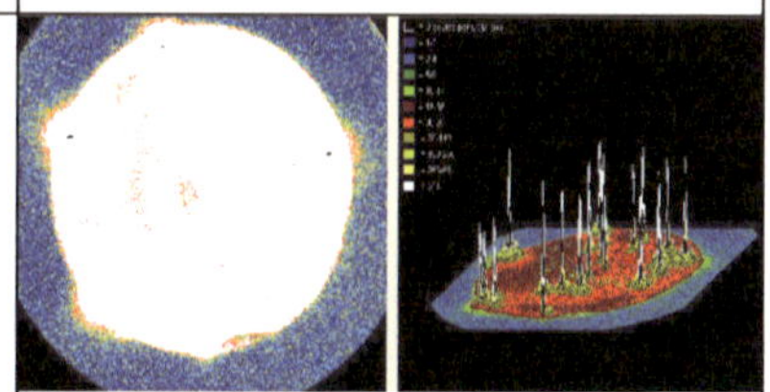
Formulation 2	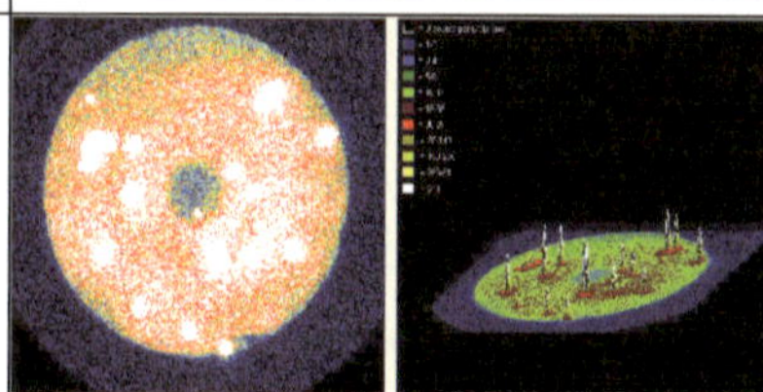	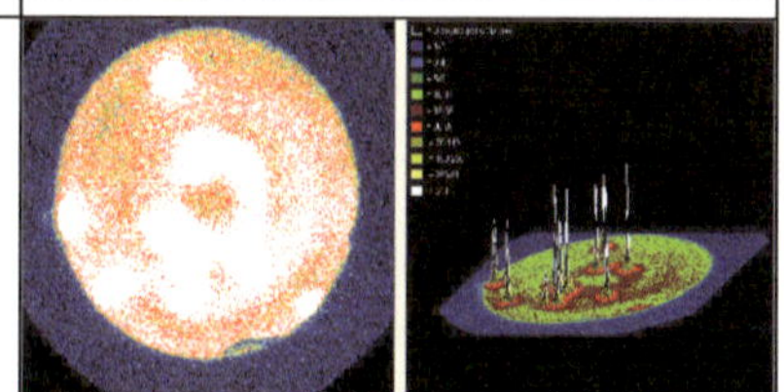
Formulation 3	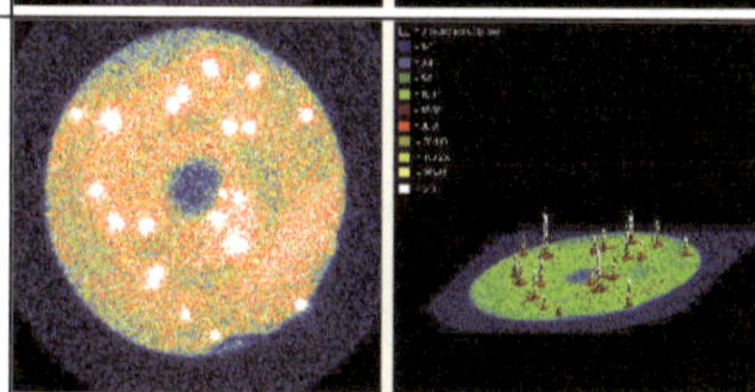	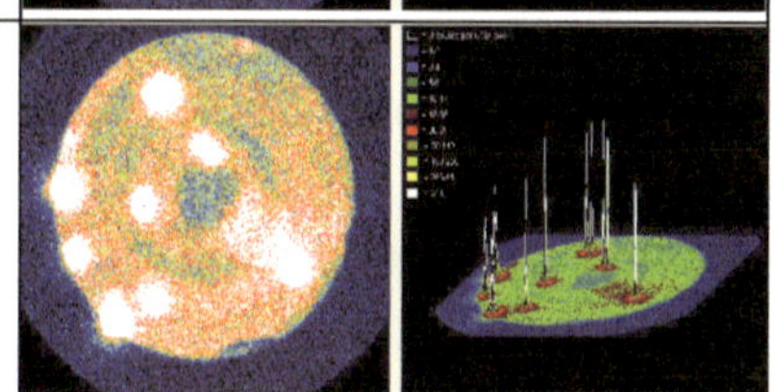
Formulation 4	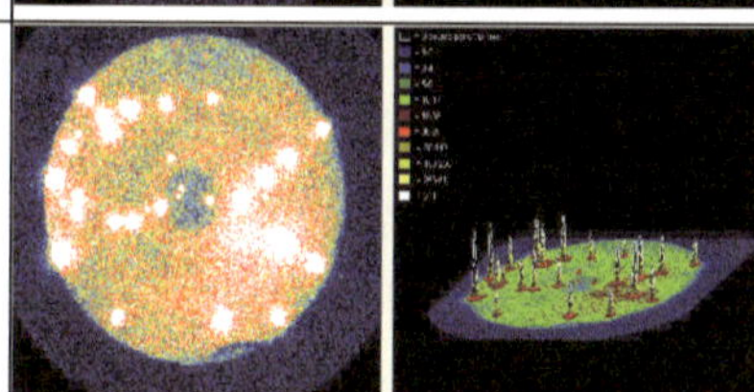	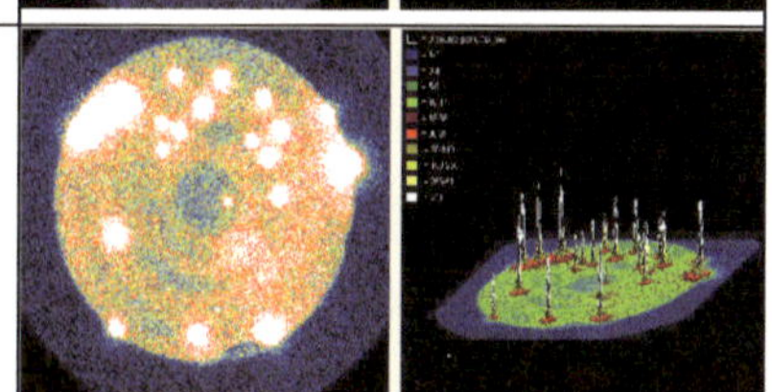
Formulation 5	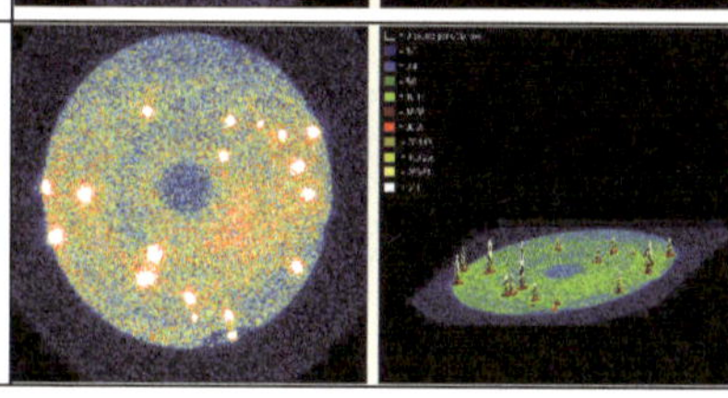	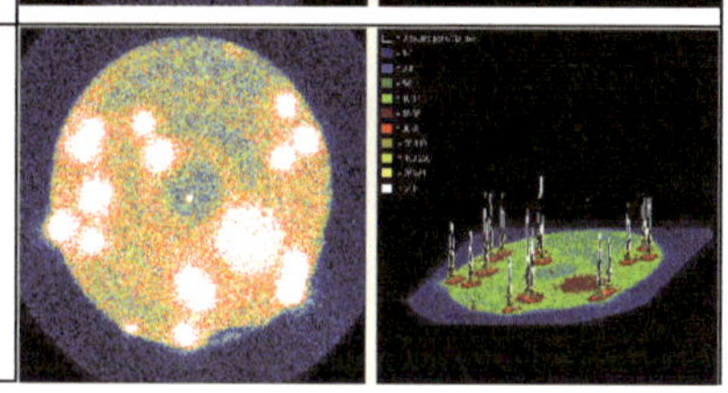

Table 2. RMDS results obtained with the 5 formulations of Luciferin/Luciferase tested

3.4 ATP background removal

One advantage to use an ATP bioluminescent assay to detect microorganisms is that ATP is present in all living organisms and is an excellent and sensitive biomarker of contamination. However this advantage can become an issue when non microbial or extracellular ATP is detected, generating bioluminescent background and preventing a reliable detection. Extracellular ATP is usually found either in culture media or in products containing eukaryotic cells. In both cases, the presence of unwanted ATP generates an overestimation of the contamination and impacts negatively the sensitivity of the ATP-bioluminescent assay. Two approaches are commonly used to remove extracellular ATP: enzymatic treatment to cleave ATP and lysis treatment to selectively lyse non bacterial cells. Methods including a treatment with ATP dephosphorylating enzymes such as apyrase or adenosine phosphatase, have been described and used to remove efficiently ATP (Askgaard et al., 1995; Thore et al., 1975). Combination of apyrase and adenosine phosphate deaminase showed a good reduction of extracellular ATP and was applied to successfully detect E. coli and *S. aureus* in media broth and biological specimens (Sakakibara et al., 1997). When the objective of the assay is to detect and quantify bacterial contamination from a mixed population containing eukaryotic cells and bacteria, a differential lysis can be applied to selectively remove eukaryotic cells from the sample. This approach was used to separate bacterial ATP from biological fluids by lysing somatic cells with detergent as Triton X 100 at low concentration and combining this step with an enzymatic degradation of ATP released from lysed cells (Chapelle et al., 1978). RMDS protocol is based on sample filtration through membrane which naturally helps to eliminate extracellular ATP. If background ATP remains after filtration, rinsing the membrane with physiological serum or sterile water contributes to removal of residual ATP and allows bacterial detection. The figure 5 shows the impact of adding rinsing steps to reduce background on beverage products.

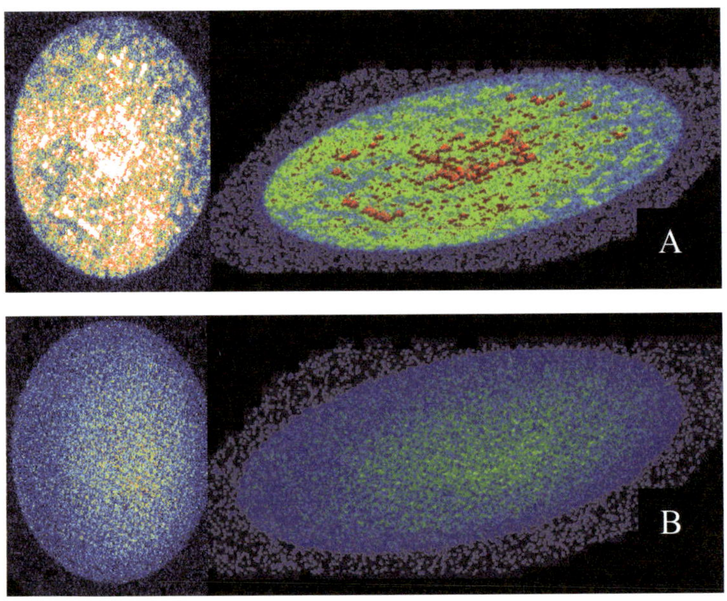

Fig. 5. Example of 2D and 3D views given by RMDS software for flavored water analysis with and without rinsing with sterile water. Picture A shows light spots corresponding to ATP present naturally in the filtered sample. Picture B shows the impact of rinsing water to remove background.

A protocol was developed to use RMDS to detect and quantify bacterial contamination from a mixture of mammalian cells and bacteria. The filtration of mammalian cells and bioluminescence detection through RMDS protocol shows (see Fig.6A) a high amount of light produced by mammalian cells preventing any bacterial detection. The sample treatment with a combination of a mammalian cells lysis solution and with apyrase contributes to efficiently remove the bioluminescent background and the figure 6B demonstrates that light spots remain detectable. These spots correspond to light emitted by bacteria in the mixture. Results obtained show that ATP-bioluminescent assay could be a powerful tool to microbiologically and quickly monitor eukaryotic cell cultures.

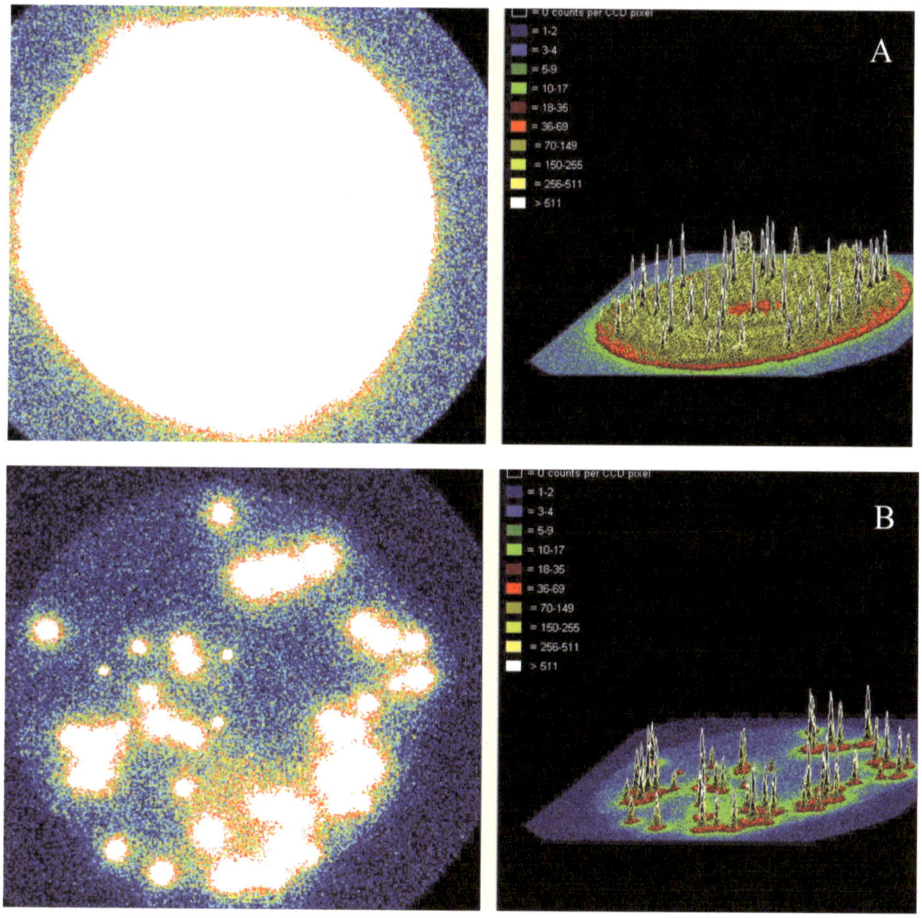

Fig. 6. A) RMDS analysis of 1mL of Chinese Hamster Ovary cells at 10^6cell/mL. Eukaryotic ATP content generates a high bioluminescent background. B) RMDS analysis of a sample containing Chinese Hamster Ovary cells at 10^6cell/mL contaminated with *E. coli* pretreated with a mammalian cells lysis solution and with apyrase. The sample pretreatment induces ATP background removal allowing contaminant detection.

3.5 RMDS applications

3.5.1 Use of Bioluminescence for microorganisms detection in water

Water is a key raw material utilized in the manufacturing of products within the food and beverage, healthcare, microelectronics and pharmaceutical industries. Within each industry, different regulatory requirements exist for microbial contamination in the water used for the manufacturing of a product for a specific application. The microorganisms found in these water systems are mainly stressed, slow-growing strains characterized by long incubation times before growth can be detected using traditional microbiology methods such as membrane filtration or pour plates. The time it takes before contamination can be detected in water can cause delays in product release, and extend the storage time of products. Using a rapid Bioluminescence based detection method allows manufacturers to identify microbial contamination earlier, which provides them with better process control, product yield, and shortens time to market.

The following table 3 provides the incubation times for detectable growth, by organism, for the traditional microbiology method and RMDS. The detection time is significantly reduced using RMDS. Detection of growth is on average 4.5 times faster than traditional microbiology, and up to 6 times faster for the very slow growers tested (*Methylobacterium mesophilicum* ATCC 29983, stressed strain of *Methylobacterium* and a mix of various slow-growing strains). RMDS allows for overnight detection of the industrial-stressed microorganisms tested. The incubation temperature also has an influence on time-to-result. Incubating at 25 °C showed that longer incubation times were required (data not shown). The mean recovery between RMDS and the traditional microbiology method in these experiments was 92.7%, which shows the equivalence of the two methods.

Microorganisms	Traditionnal Microbiology 30°C			Milliflex Rapid Detection System 30°C		
	R2A	PCA	TSA	R2A	PCA	TSA
ATCC Strains						
P. aeruginosa ATCC 9207	1 day	1 day	1 day	9 hrs	9 hrs	9 hrs
M. mesophilicum ATCC 29983	6 days	6 days	MNA	26 hrs	26 hrs	MNA
E. coli ATCC 8739	1 day	1 day	1 day	6 hrs	6 hrs	6 hrs
B. cepacia ATCC 25416	ND	ND	2 days	ND	ND	16 hrs
S. epidermidis ATCC 12228	ND	ND	1 day	ND	ND	9 hrs
Industrial-Stressed Microorganisms						
Mix of various slow-growing strains	6 days	6 days	MNA	24 hrs	24 hrs	MNA
Stressed strain of *Methylobacterium*	6 days	6 days	MNA	24 hrs	24 hrs	MNA
Environmental isolate of *R. pickettii*	2 days	2 days	ND	11 hrs	11 hrs	ND
MNA : Medium Not Appropriate for growth of microorganism, ND : Not Done						

Table 3. Detection time of reference strains and water isolates in traditional method and RMDS using either R2A agar, Tryptic Soy Agar (TSA) or Plate Count Agar (PCA).

3.5.2 Rapid detection of spores

Spores are major food spoilages and are also a concern in pharmaceutical samples. The classical microbiological method to enumerate spore contamination combines heat shock and on average 5 days incubation into sterile and molten specific medium Agar (Wayne et al., 1990). The amount of ATP in spores is very low and germination is necessary to increase ATP content and develop a rapid detection method based on ATP-bioluminescence (Kodata et al., 1996). ATP-bioluminescence rapid screening assay has been described showing that after germination, spore containing powder has been detected in a short time with a detection limit of 100 spores (Lee & Deininger, 2004). Fujinami et al (2004) also showed that short incubation of the sample in nutrient broth medium containing L-alanine increased RLU from spores and optimize the ATP- bioluminescent assay.

An easy protocol was developed to quickly enumerate spore contamination in artificially inoculated products with RMDS. Physiological water was inoculated with a calibrated concentration of *Bacillus subtilis* spores. After a heat shock at 80°C for 10 min, the inoculated product followed the protocol described in section 3.3. The incubation was performed with R2A medium at 32.5°C+/- 2.5°C. Results show that ATP bioluminescent signal start to be detected after 4h of incubation and that the reliable detection and enumeration of spores was achieved after 5 hours. Results given by RMDS are consistent with the expected inoculation level of the product and exhibit a recovery of almost 100% (tests performed in triplicate) compared with the control plate incubated 48h. Figure 7 gives an example of spore detection after 5h of incubation with R2A medium. RMDS protocol provides an alternative approach to perform rapid detection of spores in filterable matrix.

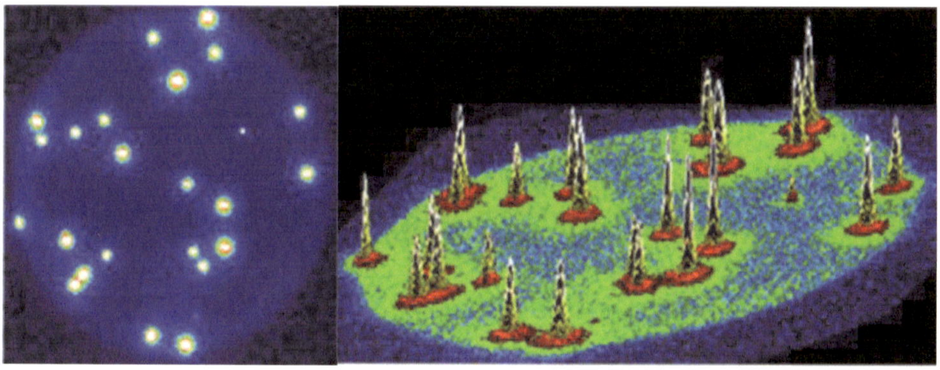

Fig. 7. RMDS 2D and 3D views showing Bacillus spore detection and enumeration.

3.5.3 Use of RMDS for the rapid detection of contaminants in bioreactor samples

RMDS has also been evaluated to detect contaminants in complex matrix containing mammalian cells. Mammalian cells including hybridoma are widely used in the biotechnology industry. Cell culture batches as well as consecutive downstream processes must be thoroughly monitored for microbial contamination. The ATP-bioluminescence technology is not selective of microbial ATP. The mammalian ATP released from the cells produces an interfering signal that must be eliminated to allow accurate counting of cfus.

Triton X100 combined with ATPase was already described to selectively extract and degrade ATP from blood products and urine samples enabling specific bacterial detection (Thore et al., 1985).

A simple and fast pretreatment method based on a selective lysis of mammalian cells and ATP removal has been developed. The harmlessness of this treatment for microorganisms was demonstrated, allowing the use of the RMDS to monitor mammalian cell samples. The protocol is as fellow: 1. Differential lysis of the mammalian cells (Chinese hamster ovary [CHO]-K1, ATCC CCL-61) using the selective mammalian cell lysis solution (Millipore MSP010053); 2. removal of mammalian ATP using 5U apyrase (Sigma Aldrich); 3. Milliflex Rapid membrane filtration using Milliflex funnel; 4. phosphate buffered saline rinsing to remove remaining mammalian ATP; 5. membrane incubation for bacteria growth; and 6. detection and counting of bacteria using the RMDS as described in the protocol.

This method enabled the detection of microorganisms in the presence of up to 5.10^7 eukaryotic cells, and involved a single pre-treatment step of the sample prior to filtration. Figure 3 (paragraphe 3.2) demonstrates that in E. coli-contaminated CHO cells, the pre-treatment removed specifically mammalian ATP and enabled the enumeration of contaminants.

The harmlessness of the cells treatment toward microorganisms was also demonstrated using *B. subtilis, S. aureus, P. aeruginosa* and *Candida albicans* spiked at approximately 50 cfus with a recovery ranging from 80% to 109.7% compared with traditional microbiology counts. During the filtration step, mycoplasma, unlike bacteria, will pass through a 0.45 µm filter (Baseman & Tully, 1997). Moreover, mycoplasma membranes are easily solubilized by detergents, and the lysis of mammalian cells simultaneously affects mycoplasma viability.

The specific mammalian cell lysis solution coupled with the RMDS method allowed fast detection of contaminating microorganisms in high value cell samples. Using RMDS to detect and quickly enumerate microbial contamination in biotechnology samples such as eukaryotic cells will allow better control throughout the process.

3.5.4 Rapid sterility testing based on ATP-Bioluminescence

In pharmaceutical companies, products are released based on microbiological quality. The Sterility test is a mandatory and critical step to ensure that the product is free of microorganism. The test takes 14 days of incubation before getting results. Time is the main reason why there is a need for an alternative and rapid method. Du to its universality and high sensitivity, the ATP-bioluminescence technology represents an alternative to ease sterility testing and shorten incubation time (Bussey & Tsuji, 1986). In addition to reduce time to detection, ATP-bioluminescence brings a solution to one drawback of current methods. Light detection replaces the subjectivity of visual determination of turbidity. Bioluminescence test that uses adenylate kinase reaction to convert ADP in ATP to significantly amplify the signal is described as a rapid sterility alternative method with results below or equal to 7 days (Albright, 2008).

Sterility testing based on RMDS follows the protocol described in section 3.2 with the major difference that the filtration step is performed under an isolator or a sterile chamber to ensure a sterile environment throughout the test. Reducing the incubation time from 14 days to 5 days is an achievable goal which benefits pharmaceutical companies. As RMDS is based on filtration, this method is compatible with complexe matrices. A comparative study was performed between RMDS and technologies based on CO2 detection. Peptone water and biological matrix such as inactivated influenza vaccines were inoculated with low concentration of microorganisms representing Gram negative, Gram positive, aerobic, anaerobic, spore forming, slow growing bacteria, yeast, and fungi. Results showed that RMDS detected all microorganisms significantly faster than the compendial method (Parveen et al., 2011). RMDS using incubation onto Schaedler Blood Agar detected all tested microorganisms in 5 days in the presence of a matrix containing preservative 0.01% thimerosal and was also compatible with inactivated influenza vaccines and aluminum phosphate or aluminum hydroxide adjuvants (Parveen et al., 2011). RMDS is likewise used as rapid sterility testing by other pharmaceutical company and shows no interference with bioluminescence mechanism and a detection in 5 days of stressed and reference strains including worst microorganism such as *Propionibacterium acnes* (Gray et al., 2010).

3.5.5 Use of RMDS for specific detection of *Pseudomonas aeruginosa*

3.5.5.1 Specific detection protocol

RMDS was used with specific hybridization probes targeting *P. aeruginosa* and coupled to Soy Bean Peroxydase. A new and unique permeabilization solution was developed and is based on polyethylimine (PEI). Cells are fixed on the membrane using a formaldehyde mix. Hybridization was performed using Peptide Nucleic Acid probes targeting 16S RNA conjugated to Soybean peroxydase diluted in Hybridization buffer. Free probes are washed with a Tween buffer. SBP catalyzes conversion of Luminol into photons and light activity of the bioluminescent reaction is recorded by the CCD camera of RMDS.

The following procedure was used to determine the minimum incubation time necessary to detect and enumerate *P. aeruginosa* with RMDS (Fig 8):

1. Pour 50 mL of saline solution into a Milliflex funnel; 2. Spike the appropriate dilution of each microorganism into the funnel (10–100 CFUs); 3.Add 50 mL of saline solution into the funnel to homogenize the content; 4. Filter and transfer the membrane onto a prefilled TSA Milliflex cassette. Incubate at 32.5 °C ± 2.5 °C for the appropriate time; 5. Once incubation is complete, separate the membrane from the cassette and let the membrane dry; 6. Follow the Milliflex Rapid *P. aeruginosa* detection procedure described before; 7. Spray the specific detection reagents using the Milliflex Rapid AutoSpray Station; 8. Read the sample with the Milliflex Rapid Detection and Enumeration System. Steps 1 through 4, 6 and 8 were performed inside a laminar flow hood.

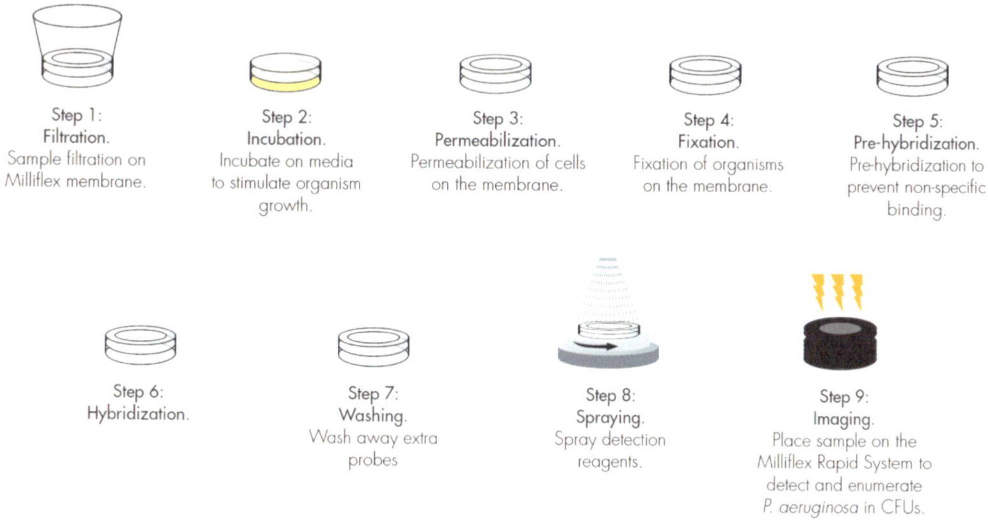

Fig. 8. *P. aeruginosa* specific detection and enumeration protocol

This procedure was also used to obtain both total viable count (TVC) and specific detection and enumeration of *P. aeruginosa* using the same membrane sample. The adapted procedure is as fellow:

pour 50 mL of saline solution into a Milliflex funnel. Spike the appropriate dilution of each microorganism into the funnel (10–100 CFUs). Add 50 mL of saline solution into the funnel to homogenize the content. Filter and transfer the membrane onto a pre-filled TSA Milliflex cassette. Incubate at 32.5 °C ± 2.5 °C for the appropriate time. Once incubation is complete, separate the membrane from the cassette and let the membrane dry. Spray the ATP releasing and bioluminescence reagents using the Milliflex Rapid AutoSpray Station. Read the sample with the RMDS. Then, follow the Milliflex Rapid P. aeruginosa detection procedure starting from fixation step. Spray the specific detection reagents using the Milliflex Rapid AutoSpray Station and read the sample with the Milliflex Rapid Detection and Enumeration System.

3.5.5.2 Specific *Pseudomonas aeruginosa* detection and total viable count results

The Milliflex Rapid system is a proven automated solution for the rapid detection and enumeration of total viable count (TVC) in purified water and Water For Injection. Based on membrane filtration and image analysis together with an adenosine triphosphate (ATP) bioluminescence reagent, the Milliflex Rapid System delivers TVC test results faster than traditional methods. We have developed a hybridization assay that enables the Milliflex Rapid system to specifically detect and enumerate *P. aeruginosa*. The hybridization assay is performed with a peroxidase-conjugated DNA-oligonucleotide probe targeted to a specific RNA-sequence of *P. aeruginosa*. Applying luminol and peroxide substrates to the membrane filtration sample generates light that is detected by the Milliflex Rapid system.

In order to determine the minimal incubation time to detect *P. aeruginosa*, a pure culture of *P. aeruginosa* ATCC 9027 was spiked into Milliflex and incubated on TSA for 6 hours for the alternative method and on R2A for 24 hours for the compendial method. Results are presented in Figure 9.

	2D view	**3D view**
Rapid Milliflex Microbiology Detection of P.aeruginosa	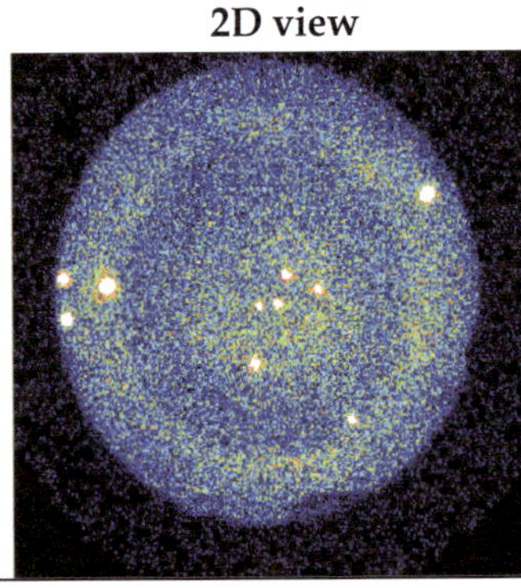	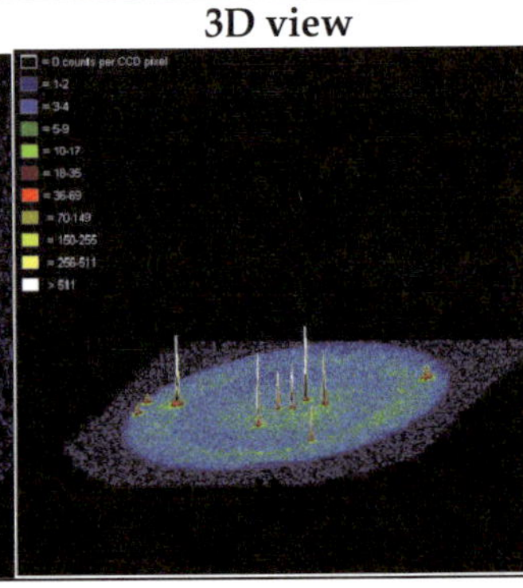
Specific Detection Count [TSA, 32.5 ± 2.5°C]	**10 CFUs**	
Traditional Microbiology Count [R2A, 25 ± 2.5°C]	**11 CFUs**	
Recovery	**91%**	
Incubation Time	**6h**	

Fig. 9. Specific Detection and enumeration of *P. aeruginosa*

Using the specific detection procedure described above *P. aeruginosa* was detected and enumerated in 8 hours in a water sample. The specificity of the method has been assessed against numerous microorganisms and only *P. aeruginosa* was detected in this panel of contaminants. The limit of the sensitivity is 1 CFU (data not shown).

The objective of this experiment was to first obtain the TVC in CFUs using the total viable count assay, followed by the specific detection assay for *P. aeruginosa*. After performing the TVC analysis, the results were stored on the Milliflex Rapid system and the same membrane was then treated following the specific detection procedure. The TVC and the specific detection count data were then analyzed (fig.10).

Figure 10 provides results for both TVC and specific detection of *P. aeruginosa* using the same membrane. The images below show that the position of each colony forming unit is identical when using the TVC and specific detection assay. One hundred percent of the CFUs were detected in each assay.

Rapid microbiology detection of TVC (A) and *P.aeruginosa* specific detection (B) using pure culture of *P.aeruginosa* ATCC 9027 [TSA, 32.5 ± 2.5°C]	**2D view** **3D view** **A** / **B** 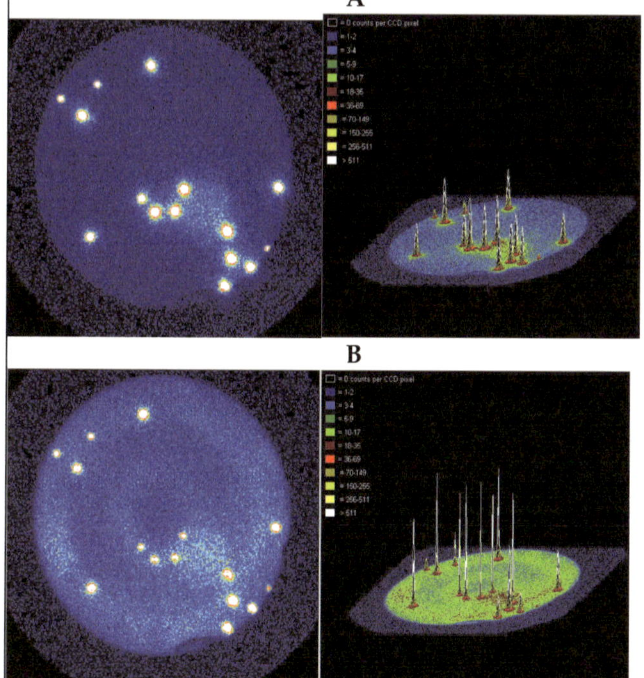
Rapid Microbiology Count	**15 CFUs**
Specific Detection Count	**15 CFUs**
Recovery	**100%**
Incubation time	**9 h**
Overall procedure	**11 h 30**

Fig. 10. Specific detection of *P .aeruginosa* after TVC on the same membrane using pure culture of *P. aeruginosa* ATCC 9027 incubated on TSA at 32.5°C+/-2.5°C.

In a second assay, a mixed microbial population composed of *P. aeruginosa, Burkholderia cepacia* and *E. coli* were spiked and analyzed with the procedure described in "Combination of Total Viable Count and Specific Detection of P. aeruginosa." Results are presented in the figure 11. After 9 hours growth at 35 °C, 24 CFUs were detected after the TVC procedure and 8 CFUs were detected using the *P. aeruginosa* specific detection procedure.

This demonstrates that the system is able to make TVC and specific detection even in a mixed population of microorganisms.

Milliflex Rapid Microbiology Detection of TVC (A) and *P.aeruginosa* specific detection (B) in a mixed population composed of *P.aeruginosa*, *B.cepacia* & *E. coli* [TSA, 32.5 ± 2.5°C]	**2D view** **3D view** 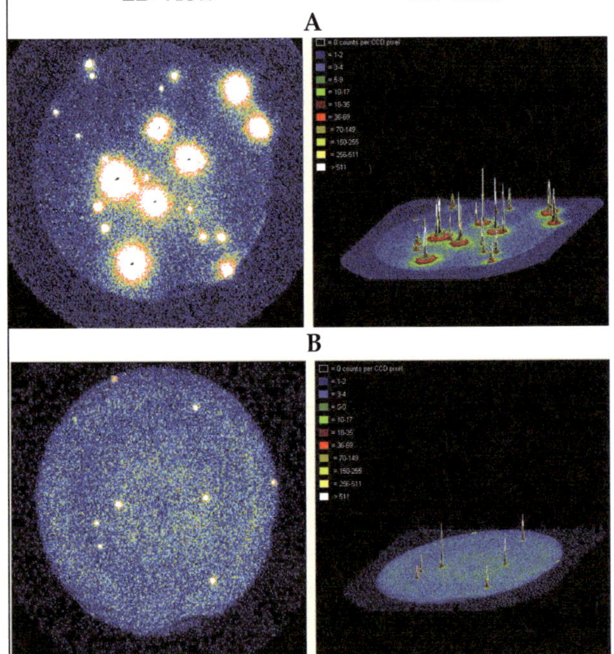
Rapid microbiology count	**24 CFUs**
Specific detection count	**8 CFUs**
Percentage of *P.aeruginosa* contaminants in the mix	**29%**
Incubation Time	**9 h**
Overall procedure	**11 h 30**

Fig. 11. Specific detection of *P. aeruginosa* after TVC on the same membrane using a mixed population of *P. aeruginosa* ATCC 9027, B. cepacia ATCC 25416 and E.coli ATCC 25922 incubated on TSA at 32.5°C+/-2.5°C.

4. Conclusion

The different studies presented here show how versatile is the use of Bioluminescence for microorganisms detection. We demonstrate here that it offers a high sensitivity to detect microbial contamination rapidly in a variety of filterable samples.

The association of Bioluminescence to sensitive sensors such as RMDS provides a result in colony forming units equivalent to the standard plate count but is 4 times faster than classical microbiology. This method can be used in samples from industrial water, to food and beverage samples for the detection of any type of bacteria, yeasts and molds including spores. We also showed that it can be used to detect bacterial contamination in cell culture matrices containing high concentrations of eukaryotic cells.

Interestingly, Bioluminescence was also coupled to molecular biology through the use of 16S RNA probes for specific detection of bacteria. The example presented here allowed not only the detection of *P. aeruginosa* but also the total viable count using Luciferin and luciferase followed by specific detection of this very specific bacterium.

Finally, the development of the method in a pharmaceutical environment allowed sterility testing of drug products 3 times faster than the compendial method. This recent developments in the pharmaceutical field show that the method is also able to help patients taking drugs usually associated with a very short shelf life (gene therapy products, cell therapies...) as the result is delivered before the injection of the product while the traditional systems usually deliver after the treatment.

In conclusion, the use of Bioluminescence either in its "classical" or molecular format allows for a number of developments in the field of microorganisms detection. The flexibility of the method and its ease of use coupled to the considerable savings in time compared to the traditional method make it a valuable tool for life scientists as well as for other clinical applications.

5. Acknowledgment

Authors would like to thanks colleagues from Merck-Millipore Application group, Development group and Predevelopment - Technology – Collaboration for their technical collaboration. The research described in this paper was carried out at the Merck-Millipore R&D laboratory (Molsheim, France).

6. References

Albright, J. (2009). Implementing Rapid Sterility using the Celsis Enhanced ATP Bioluminescence Test. www.celsis.com/media/pdf/rdpdfs/Poster_RapidSterilityTesting_PDA0904.pdf

Andreotti, P. E. & Berthold, F. (1999). Application of a new high sensitivity luminometer for industrial microbiology and molecular biology. *Luminescence*, 14(1), 19-22.

Askgaard, D. S.; Gottschau, A; Knudsen, K. & Bennedsen, J. (1995). Firefly luciferase assay of adenosine triphosphate as a tool of quantitation of the viability of BCG vaccines. *Biologicals*, 23(1), 55-60.

Aycicek, K., Oguz, U. & Karci, K. (2006). Comparison of results of ATP bioluminescence and traditional hygiene swabbing methods for the determination of surface cleanliness at a hospital kitchen. *International Journal of Hygiene and Environmental Health*, 209(2), 203-206.

Baseman, J. B. & Tully, J.G. (1997). Mycoplasmas: Sophisticated, reemerging and burdened by their notoriety. *Emerging Infectious Disease*, 3, 21-32.

Bautisda, D. A.; Vaillancourt, J. P., Clarke, R. A. ; Renwick S. & Griffiths M. W. (1995). Rapid assessment of the microbiological quality of poultry carcasses using ATP-bioluminescence. *Journal of Food Protection*, 58, 551-554.

Branchini, B. R.; Murtiashaw, M. H.; Magyar, R. A. & Anderson, S. M. (2000). The role of lysine 529, a conserved residue of the acyl-adenylate-forming enzyme superfamily, in firefly luciferase. *Biochemistry*, 39, 5433–5440.

Branchini, B. R.; Southworth, T. L.; Khattak, N. F.; Michelini, E. & Roda, A. (2005). Red- and green-emitting firefly luciferase mutants for bioluminescent reporter applications. *Analytical Biochemistry*, 345, 140–148.

Bussey, D. M. & Tsuji, K. (1986). Bioluminescence for USP sterility testing of pharmaceutical suspension products. *Applied Environmental Microbiology*, 51(2), 349-355.

Chapelle, E. W.; Picciolo, G.L. & Deming, J.W. (1978). Determination of bacterial contents in fluid. *Methods in Enzymology*. 57, 65-72.

Chen, F. C. & Godwin, S. L. (2006). Comparison of a rapid ATP bioluminescence assay and standard plate count methods for assessing microbial contamination of consumers' refrigerators. *Journal of Food Protection®*, 69(10), 2534-2538.

Conn, R. B.; Charache, P. & Chapelle, E. W. (1975). Limits of applicability of the firefly luminescence ATP assay for the detection of bacteria in clinical specimens. *American Journal of Clinical Pathology*, 63, 493-501.

Conti, E.; Franks, N. P. & Brick, P. (1996) Crystal structure of firefly luciferase throws light on a superfamily of adenylate-forming enzymes. Structure, 4, 287–298.

Davidson, C. A.; Griffith, C. J.; Peters, A. C. & Fielding, L. M. (1999). Evaluation of two methods for monitoring surface cleanliness-ATP bioluminescence and traditional hygiene swabbing. *Luminescence*, 14(1), 33-38.

DeLuca, M. & McElroy, W. D. (1974). Kinetics of the firefly luciferase catalyzed reactions. *Biochemistry*, 13, 921–925.

Deininger, R. A. & Lee, J. Y. (2001). Rapid determination of bacteria in drinking water using an ATP assay. *Field Analytical Chemistry & Technology*, 5(4), 185-189.

Dostalek, P. & Branyik T. (2005). Prospects for rapid bioluminescent detection methods in the food industry - a review. *Czech Journal of Food Sciences*, 23(3), 85-92.

Francis, K. P.; Joh, D.; Bellinger-Kawahara, C.; Hawkinson, M. J.; Purchio, T. F. & Contag, P. R. (2000). Monitoring bioluminescent Staphylococcus aureus infections in living mice using a novel luxABCDE construct. *Infection. Immunity*, 68, 3594–3600.

Frundzhyan, V. & Ugarova, N. (2007). Bioluminescent assay of total bacterial contamination of drinking water. *Luminescence*, 22(3); 241-244.

Fujinami, Y.; Kataoka, M.; Matsushita, K.; Sekigushi, H.; Itoi, T.; Tsuge, K. & Seto, Y. (2004). Sensitive Detection of Bacteria and Spores Using a Portable Bioluminescence ATP Measurement. Assay System Distinguishing from White Powder Materials. *Journal of Health science*, 50, 126-132.

Girotti, S.; Ferri, E.N.; Fini, F.; Righetti, S.; Bolelli, L.; Budini, R.; Lasi, G.; Roubal, P.; Fukal, L.; Hochel, I. & Rauch, P. (1997). Determination of microbial contamination in milk by ATP assay. *Czech Journal of Food Science*, 15, 241-248.

Gray, J. C.; Steark, A.; Berchtold, M.; Mercier, M.; Neuhaus, G. & Wirth A. (2010). Introduction of a Rapid Microbiological Method as an Alternative to the Pharmacopoeial Method for the Sterility test. *American Pharmaceutical Review*. 13(6), 88-94.

Hawronskyj, J. M. & Holah, J. (1997) ATP: a universal hygiene monitor. *Trends in Food Science & Technology*, 8, 79–84.

Hosseinkhani, S. (2011). Molecular enigma of multicolor bioluminescence of firefly luciferase. *Cellular and Molecular Life Sciences*, 68(7),1167-1182.

Kodaka, H.; Fukuda, K.; Mizuochi, S. & Horigome, K. (1996). Adenosine Triphosphate Content of Microorganisms Related with food Spoilage. *Japanese Journal of Food Microbiology*, 13, 29-34.

Kolbeck, J. C.; Padgett, R. A.; Estevez, E. G. & Harell, L. J. (1985). Bioluminescence screening for bacteriuria. *Journal of clinical Microbiology*, 21, 527-530.

McElroy, W. D. (1947). The energy source for bioluminescence in an isolated system. *Proceedings of the National Academy of Sciences USA*, 33, 342–345.

McElroy, W. D. (1951). Properties of the reaction utilizing adenosinetriphosphate for bioluminescence. *Journal of Biological Chemistry*, 191, 547–557.

McElroy, W., D., Hastings, J., W.; Coulombre, J. & Sonnenfeld, V. (1953). The mechanism of action of pyrophosphate in firefly luminescence. *Archives of Biochemistry and Biophysics*, 46, 399–416.

Miller, J. N.; Nawawi, M. B. & Burgess, C. (1992). Detection of bacterial ATP by reversed flow-injection analysis with luminescence detection. *Analytica Chimica Acta*, 266, 339-343.

Nakatsu, T.; Ichiyama, S.; Hiratake, J.; Saldanha, A.; Kobashi, N.; Sakata, K. & Kato, H. (2006). Structural basis for the spectral difference in luciferase bioluminescence. *Nature*, 440, 372–376.

Nielsen, P. & Van Dellen, E. J. (1989). Rapid bacteriological screening of cosmetic raw materials by using bioluminescence. *Association of Official Analytical Chemists*, 72(5), 708-711.

Parveen S.; Kaur S.; David S. A.; Kenney J. L.; McCormick W. M. & Gupta R.K. (2011). Evaluation of growth based rapid microbiological methods for sterility testing of vaccines and other biological products. *Vaccine.*

Poulis, J. A. ; Phper, M. & Mossel, D. A. A. (1993). Assessment of cleaning and desinfection in the food industry with the rapid ATP-bioluminescence technique combined with the tissue fluid contamination test and a conventional microbiological method. *International Journal of food Microbiology*, 20, 109-116.

Roda, A.; Pasini, P.; Mirasoli, M.; Michelini, E. & M. Guardigli. (2004). Biotechnological applications of bioluminescence and chemiluminescence. *Trends in Biotechnology*, 22, 295-303.

Sakakibara, T.; Murakami, S.; Hattori, N.; Nakajima, M. & Imai, K. (1997). Enzymatic treatment to eliminate the extracellular ATP for improving the detectability of bacterial intracellular ATP. *Analytical Biochemistry*, 250(2), 157-161.

Selan, L.; Berlutti, F.; Passariello, C.; Thaller, M. C. & Renzini, G. (1992). Reliability of a bioluminescence ATP assay for detection of bacteria. *Journal of Clinical Microbiology*, 30, 1739-1742.

Seliger, H. H. (1989). Some reflections on McElroy and bioluminescence. *Journal of Bioluminescence and Chemiluminescence*, 4(1), 26-28.

Seliger, H. H.; Buck, J. B.; Fastie, W. G. & McElroy, W. D. (1964). The spectral distribution of firefly light. *Journal of General Physiology*, 48, 95–104.

Seliger, H. H. & McElroy, W. D. (1964). The colors of firefly bioluminescence: enzyme configuration and species specificity. *Proceedings of the National Academy of Sciences USA*, 52, 75–81.

Shapiro, E.; Lu, C. & Baneyx, F. (2005). A set of multicolored Photinus pyralis luciferase mutants for in vivo bioluminescence applications. *PEDS* 18 (12), 581-587.

Siro, M-R.; Romar, H. & Lövgren, T. (1982). Continuous flow method for extraction and bioluminescence assay of ATP in baker's yeast. *Applied Microbiology and Biotechnology*, 15, 258-264.

Thompson, J. F.; Geoghegan, K. F.; Lloyd, D. B.; Lanzetti, A. J.; Magyar, R. A.; Anderson, S. M., & Branchini, B. R. (1997). Mutation of a protease-sensitive region in firefly luciferase alters light emission properties. *Journal of Biological Chemistry*, 272, 18766–18771.

Thorne, N; Inglese, J. & Auld, D. S. (2010). Illuminating insights into firefly luciferase and other bioluminescent reporters used in chemical biology. *Chemistry & Biology*, 17(6), 646-657.

Thore, A.; Ansehn,S.; Lundin, A. & Bergman, S. (1975). Detection of bacteria by luciferase assay of adenosine triphosphate. *Journal of Clinical Microbioliology*, 1, 1-8.

Venkateswaran, K.; Hattori, N.; La Duc, M.T. & Kern, R. (2003). ATP as a biomarker of viable microorganisms in clean-room facilities. *Journal of Microbiological Methods*, 52(3), 367-377.

Wilson, T. & Hastings, J.W. (1998). Bioluminescence. *Annual Review of Cell and Developmental biology*, 14, 197-230.

Yan, S. L.; Miao, S. N.; Deng, S. Y.; Zou, M. J.; Zhong, F. S.; Huang, W. B.; Pan, S. Y. & Wang, Q. Z. (2011). ATP bioluminescence rapid detection of total viable count in soy sauce. *Luminescence.*

Zako, T.; Ayabe, K.; Aburatani, T.; Kamiya, N.; Kitayama, A.; Ueda, H. & Nagamune, T. (2003). Luminescent and substrate binding activities of firefly luciferase N-terminal domain. *Biochimica et Biophysica Acta - Proteins & Proteomics*, 1649, 183–189.

Development of a pH-Tolerant Thermostable *Photinus pyralis* Luciferase for Brighter *In Vivo* Imaging

Amit Jathoul[1], Erica Law[2], Olga Gandelman[3,*],
Martin Pule[1], Laurence Tisi[3] and Jim Murray[4]
[1]*Cancer Institute, University College London,*
[2]*Illumina Inc., Chesterford Research Park,*
[3]*Lumora Ltd., Cambridgeshire Business Park,*
[4]*School of Biosciences, Cardiff University*
UK

1. Introduction

Firefly luciferase (Fluc) catalyzes a bioluminescent reaction using the substrates ATP and beetle luciferin in the presence of molecular oxygen (Fig. 1A). Because of its use of ATP and the simplicity of the single-enzyme system, firefly luciferase is widely used in numerous applications, notably those involving detection of living organisms, gene expression or amplification in both *in vivo* and *in vitro* systems.

$$LH_2 + ATP + O_2 \rightarrow LO + AMP + PP_i + CO_2 + h\nu$$

A

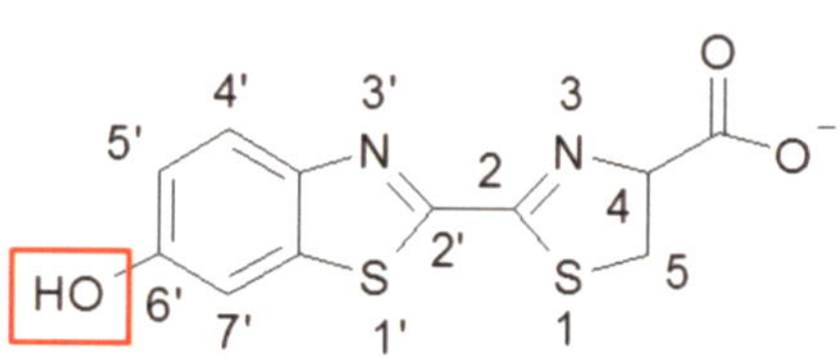

B

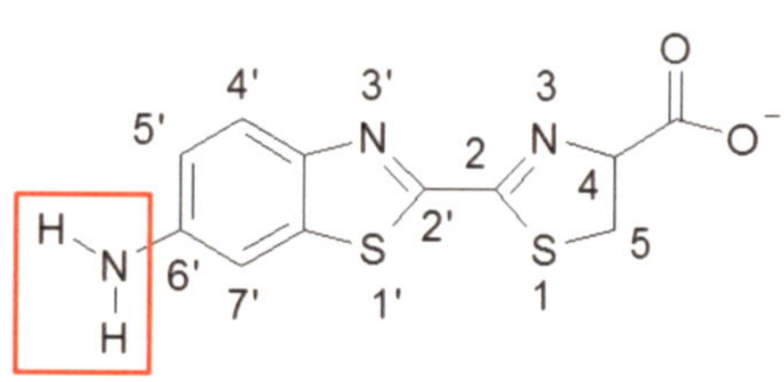

C

Fig. 1. Bioluminescent reaction of Fluc (A) and chemical structures of luciferin (LH_2) (B) and aminoluciferin (ALH_2) (C) eliciting intense bioluminescence.

Fluc has found intensive application in small animal *in vivo* bioluminescence imaging (BLI) in which the activity or state of labelled proteins, cells, tissues and organs may be localised

*Corresponding Author

and quantified sensitively and non-invasively. Different models and procedures for BLI are well described (Kung, 2005; Zinn *et al.*, 2008). For example, BLI is routinely applied to serially detect the burden of xenografted tumours in mice. Using more complex techniques, such as Fluc re-complementation, protein interactions such as chemokine receptor dimerisation (Luker *et al.*, 2008) have been imaged in small animals.

Recently, BLI has also been adapted to the detection of small molecules *in vivo* (Van de Bittner *et al.*, 2010). *D*-LH_2 is typically given intravenously or intraperitoneally to mice and has a broad biodistribution profile (Berger *et al.*, 2010). Cellular levels of Mg and ATP are sufficient to drive the reaction, though kinetics depends on substrate diffusion. Light emitted from labelled cells is detected using imagers which consist of CCD cameras in a dark box. This gives invaluable insight into the effects of experiments in the context of living organisms in real time.

The advantages of BLI over comparative techniques such as positron emission tomography (PET) include its simplicity, low cost, non-requirement for radiation and versatility. The other main optical imaging technique, fluorescence imaging (FLI), in which fluorescent small molecules or proteins are imaged in small animals, has lower signal to noise ratio than BLI due to the background signal in FLI from autofluorescence, quenching of signal due to endogenous tissue chromophores and also the requirement, and dependence on penetration, of an excitation light. Thus, BLI is approximately three orders of magnitude more sensitive than FLI and has a very large dynamic range (Wood, 1998).

All optical imaging techniques suffer from low resolution and from wavelength dependence of imaging due to photon scatter and signal attenuation by endogenous absorbing compounds. For example haemoglobin absorbs strongly below 590 nm. Therefore it is the red part of the spectra that is detected most efficiently (Caysa *et al.*, 2009).

Wild-type (WT) luciferase is highly thermolabile, inactivating and bathochromic shifting at even room temperature, and is sensitive to buffer conditions such as pH (Law *et al.*, 2006). The recombinant WT luciferase retains between 30 and 45% of activity at pH 7.0 relative to that at the optimal pH of 7.8 - 8.0, depending on whether flash heights or integrated light were measured (Law *et al.*, 2006). While many *in vitro* applications, such as those used to detect DNA amplification (Gandelman *et al.*, 2010) or ATP-assays (Strehler, 1968), take place at alkaline pH values of 8.0 or higher, *in vivo* applications such as medical or whole cell imaging must take place at neutral pH of 6.9 - 7.2, dependent on the exact cell type. At 37°C, WT Fluc shows a bathochromic shift and thus emits predominantly red light. Though light of red and longer wavelength does penetrate tissues more readily, red-shift luciferase bioluminescence is usually accompanied by a significant reduction in quantum yield and is therefore as such undesirable (Seliger and McElroy, 1959; Seliger and McElroy, 1960) as fluctuating levels of luciferase activity make quantitative studies problematic. Furthermore, for multispectral purposes (Mezzanotte *et al.*, 2011), any shift is undesirable. Therefore thermostable enzymes of different colours, which resist bathochromic shift are preferable.

The ideal Fluc would be highly thermostable, resist bathochromic shift, bright, have favourable kinetics (such as high substrate affinity) and have increased pH-tolerance. To address the issues of thermostability and bathochromic shift a number of recombinant mutant luciferases with increased thermal stability, giving brighter and more stable signals at elevated temperatures and resistant to bathochromic shift have been developed using protein engineering (Hall *et al.*, 1999; Tisi *et al.*, 2002; Branchini *et al.*, 2009). Enhanced thermostability greatly improves the brightness achievable *in vivo*, and such enzymes are just recently finding application in animals (Law *et al.*, 2006; Baggett *et al.*, 2004; Mezzanotte *et al.*, 2011; Michelini *et al.*, 2008). A combination of higher thermal stability with increased pH-tolerance of Fluc is a very much desired and favourable feature for *in vivo* imaging which is likely to be useful in other applications (Foucault *et al.*, 2010).

Thermostability can be greatly improved by one amino acid change of FLuc and it has been observed that both changes in the enzyme core and on the protein surface can alter stability (Tisi *et al.*, 2002b). The substitutions A217I, L or V, identified by random mutagenesis increase the thermo- and pH-stability of *Luciola cruciata* and *L. lateralis* Flucs, and the equivalent substitution in *Photinus pyralis* (*Ppy*) Luc (A215L) also increases thermostability (Kajiyama and Nakano, 1993; Squirrell *et al.*, 1998). By random mutagenesis of *Ppy* Luc, and N-terminal surface loop-based substitutions, E354K or E354R have been identified to increase thermostability (White *et al.*, 1996). Combination of E354 with mutation D357 produced thermostable double mutants, of which E354I/ D357Y (x2 Fluc; Table 1) and E354R/ D357F were shown to be more stable than D357Y or E354K alone (Willey *et al.*, 2001).

Cumulative addition of such mutations further enhances thermostability. A typical example is a mutant containing T214C, I232A, F295L and E354K, named x4 Luc (Table 1) (Tisi *et al.*, 2002b). Non-conserved surface-exposed hydrophobic residues previously mutated to Ala (Tisi *et al.*, 2001; Prebble *et al.*, 2001) have also been substituted for polar ones (F14R, L35Q, V182K, I232K and F465R) to produce a mutant, named x5 Luc, displaying additively improved thermostability, solvent stability and pH-tolerance in terms of activity and resistance to red-shift; while retaining the same specific activity relative to WT luciferase (Law *et al.*, 2002; Law *et al.*, 2006). The most thermostable mutant luciferase, Ultra-Glo™ (UG) was created from *Photuris pennsylvanica* luciferase, and is commercially available for a number of assays (Hall *et al.*, 1999; Woodroofe *et al.*, 2008).

Majority of studies on improving thermostability, pH-tolerance and brightness of Fluc have been carried out using LH_2 until now. Emerging applications of luciferases in *in vivo* imaging of protease activity (Dragulescu-Andrasi *et al.*, 2009) require a different substrate - ALH_2, one of the very few LH_2 analogues with which firefly luciferase also produces bioluminescence of relatively high intensity (White *et al.*, 1966). The substitution of the 6'-group extends the range of groups that can be conjugated to luciferin, for example to amino acids (Shinde *et al.*, 2006), peptides (Monsees *et al.*, 1995) and linear or bulky N-alkyl groups (Woodroofe *et al.*, 2008). Peptide-conjugated pro-luciferins allow the bioluminescent measurement of protease activity and in such applications ALH_2 (Monsees *et al.*, 1995) therefore the properties of different firefly luciferases and their mutants with ALH_2 may impact on the choice of enzymes applied.

The limited data on *Ppy* Fluc bioluminescence with ALH_2 as a substrate show that these properties are very different from those of LH_2. The bioluminescence colour with ALH_2 has long been reported as pH-independent orange-red (max 605nm) (White *et al.*, 1966), Km for ALH_2 is approximately 26-times lower and Vmax is 10 times lower than that of LH_2 (Shinde *et al.*, 2006). There has been no further analysis of either red-shifted excited state emitter with

ALH_2 or its higher catalytic efficiency. From other studies it is known that there are luciferase isoforms from *Pyrophorus plagiopthalamus* that emit green light (*PpldGr*: 550 nm at pH 7.6) and yellow light (*PplvY*: 577 nm at pH 7.6) with ALH_2. This indicates that red emission is not an intrinsic property of ALH_2, but merely a consequence of enzymatic interactions and conformation of the active site (White *et al.*, 1966; Nakatsu et al., 2006; Branchini *et al.*, 2001; Sandalova and Ugarova, 1999). Thus, it should be possible to engineer luciferase mutants with advantageous properties with ALH_2, such as altered emission colour, higher activity and/or kinetics beneficial for *in vivo* imaging of protease activity.

In this paper we report on the construction and characterisation of a further improved x12 mutant based on the x5 mutant and seven additional mutations. Each of these mutations has previously been shown to confer slower rates of thermal inactivation (White *et al.*, 1996; Squirrell *et al.*, 1999; Tisi *et al.*, 2002). We compared the performance of x12 mutant with the WT *Ppy* and UG luciferases and demonstrated its pH tolerance and increased thermostability. A reversion of one of the mutations in the x12 resulted in a simplified mutant, termed x11 Fluc. Herein, we present properties of this mutant, which is highly thermostable, pH-tolerant, has high activity and catalytic efficiency with both LH_2 and ALH_2, and presents a great potential for *in vivo* applications with both substrates.

2. Materials and methods

2.1 Materials

D-LH_2 potassium salt was obtained from Europa Bioproducts and *D*-ALH_2 from Marker Gene Technologies Inc. (Eugene, OR, USA). x2 Fluc was donated by Dr. Peter White (Dstl, Porton Down, Salisbury, UK) [White *et al.*, 2002; Willey *et al.*, 2001]. Ultra-Glo™ luciferase (UG) was purchased from Promega and all other chemicals were purchased from Melford Laboratories Ltd. or Sigma-Aldrich unless otherwise specified.

2.2 Construction of the x12 Fluc mutant and revertants

Seven mutations were introduced sequentially onto the ×5 luciferase gene in pET16b-luc×5 (Law *et al.*, 2006) using the QuickChange™ Site Directed Mutagenesis (SDM) kit (Stratagene) according to the manufacturer's protocol. The primers used for the seven rounds of SDM are as follow:

5′-GCAGTTGCGCCC**GTG**AACGAC-3′ and 5′-GTCGTT**CAC**GGGCGCAACTGC-3′ for A105V; 5′-CCCTATTTTCATTC**CTG**GCCAAAAGCACTC-3′ and 5′-GAGTGCTTTTGGC**CAG**GAATGAAAATAGGG-3′ for F295L; 5′-GGCTACAT**ACT**GGAGACATAGC-3′ and 5′-GCTATGTCTCC**AGT**ATGTAGCC-3′ for S420T;

5′-CAAATCAAACCG**GGT**ACTGCGATTTTAAG-3′ and 5′-CTTAAAATCGCAGT**ACC**CGGTTTGATTTG-3′ for D234G; 5′-CCGCATAGA**TGT**GCCTGCGTCAGATTC-3′ and 5′-GAATCTGACGCAGGC**ACA**TCTATGCGG-3′ for T214C;

5′-CA<u>CCC**CGC**</u>GGGGAT**TAT**AAACCGGG-3′ and 5′-CCCGGTTT**ATA**ATCCCC<u>**GCG**GGG</u>TG-3′ (AvaI) for E354R and D357Y

Boldface type represents the mutated codon, underlined letters represent modified endonuclease site used to facilitate screening, and the endonuclease used for screening is shown in parentheses. *E. coli* BL21 (pLysS) (Edge Biosystems, Gaithersburg, MD, USA or XL2-Blue ultracompetent cells (Stratagene) were used as cloning hosts for the generation and selection of mutants from site-directed mutagenesis. Expression from colonies was induced by adsorbing colonies onto Hybond™-N nitrocellulose membranes (Amersham Biosciences Corp., Piscataway, NJ, USA) and transferring membranes onto fresh Luria Bertani (LB) agar plates containing 100 µg/ml carbenicillin and 1 mM IPTG and incubating for 3 hours at room temperature (RT). Bioluminescence was initiated by spraying membranes with 1 mM LH_2 or 500 µM ALH_2 in 0.1 M citrate buffer (pH 5) and colony screening was carried out by photographing emitted light with Nikon D70S camera (Nikon Corp., Tokyo, Japan). After seven rounds of SDM, mutations introduced were confirmed by sequencing of the entire luciferase gene using a facility provided by the Department of Genetics, University of Cambridge.

2.3 Expression and purification of ×12 Fluc and revertants

His_{10}-tagged WT recombinant luciferase (WT) and mutants were expressed and purified according to the optimised protocol described in (Law *et al.*, 2006). Total protein concentrations were estimated by the method of Bradford (Bradford, 1976), using the Coomassie Blue protein assay reagent kit from Pierce according to the manufacturer's protocol, with BSA as the standard.

2.4 Luciferase activity assays, kinetic analysis, pH dependence of activity, thermal inactivation and bioluminescence spectra

Luciferase mutants were diluted from purified stock solutions into pre-chilled 0.1 M Tris/ acetate; pH 7.8, 2 mM EDTA and 10 mM $MgSO_4$ (TEM) containing 2 mM DTT to obtain the required concentration, unless specified otherwise. Refer to caption accompanying each table or figure for method details. Bioluminescence spectra were captured using a Varian fluorometer (Palo Alto, Ca, USA). For measurements at differing pH values, TEM buffer at different pH values was used to dilute substrates and enzymes. Data were corrected for variant PMT sensitivity as previously described (Law *et al.*, 2006).

2.5 Mammalian cell culture, retrovirus production and transduction of cells

The genes encoding WT Fluc and x11 Fluc from pET16b constructs were cloned into mammalian retroviral expression vector SFG fused to Myc tags. These constructs were triple transfected into 293T cells, cultured in IMDM (Lonza, Basel, Switzerland) with 10% fetal calf serum (FCS) (Hyclone Labs Inc., Logan, UT, USA) and 1 % glutamax (Invitrogen Corp., Groningen, The Netherlands), along with plasmids encoding retroviral envelope and gagpol genes to produce retrovirus, which was used to transduce Raji cells. Transduced cells were sorted by flow cytometry using a Moflo-XDP instrument (Beckman Coulter, CA, USA) by anti-myc.FITC (Santa Cruz Biotechnology Inc., CA, USA) staining of the same mean fluorescence intensity and were cultured in RPMI 1640 (Lonza, Basel, Switzerland) with 10% FCS and 1 % glutamax in 5% CO_2.

2.6 *In vivo* imaging

Three month old Beta2m-mice were tail vein injected with 1 x 10^6 Raji cells expressing WT or x11 Fluc. These were imaged using 30-60 s exposures in an IVIS 200 imager (Caliper, NJ, USA) at days 3 and 10 after anaesthesia with isofluorane and 15 min after intra-peritoneal (i.p.) injection of 200 μl of sterile D-luciferin (Regis Technologies, IL, USA).

3. Results and discussion

3.1 Construction, expression, purification and thermal inactivation of ×12 Fluc and subset mutants

Seven additional mutations, previously shown to confer slower rates of thermal inactivation of Fluc, were sequentially added onto the ×5 Fluc by SDM to create x12 Fluc, which was expressed in BL21(DE3)pLysS and purified to > 90 % homogeneity as previously described (Law *et al.*, 2006; White *et al.*, 1996; Tisi *et al.*, 2002; Squirrell *et al.*, 1999).

Mutation	Mutants					Location
	x12	x11	x5	x4	x2	
F14R	+	+	+			Surface
L35Q	+	+	+			Internal
A105V	+	+				Surface
V182K	+	+	+			Surface
T214A				+		Internal
T214C	+	+				Internal
I232A				+		Surface
I232K	+	+	+			Surface
D234G	+	+				Surface
F295L	+			+		Internal
E354R	+	+			+	Surface
E354K				+		Surface
D357Y	+	+			+	Surface
S420T	+	+				Surface
F465R	+	+	+			Surface

Table 1. Mutations and their positions in thermostable Fluc mutants

The loss of activity of ×12 Fluc was measured at 55°C in two buffers, one of which allows direct comparison with the previously described ×5 Fluc (Fig. 2A - buffer A) and the other, mimicking conditions used in BART (bioluminescent assay for monitoring nucleic acid amplification in real-time) (Fig. 2A - buffer B) (Tisi *et al.*, 2002) . ×12 Fluc retained 80% of starting total activity after 30 min of treatment at 55°C, whereas ×5 Fluc had < 1% activity remaining after 5 min at the same temperature (results not shown). When compared to previous thermostable Fluc mutants (Table 1) containing subsets of x12 Fluc mutations, x12 Fluc was the most resistant to thermal inactivation at 40°C (85-90% of initial activity after 1hr) (Fig. 2B), followed by x4 and x2 Fluc (both 75-80% after 1hr) and then x5 Fluc (20% after 1hr), which are all more stable than WT Fluc (fully inactivated within 10 min).

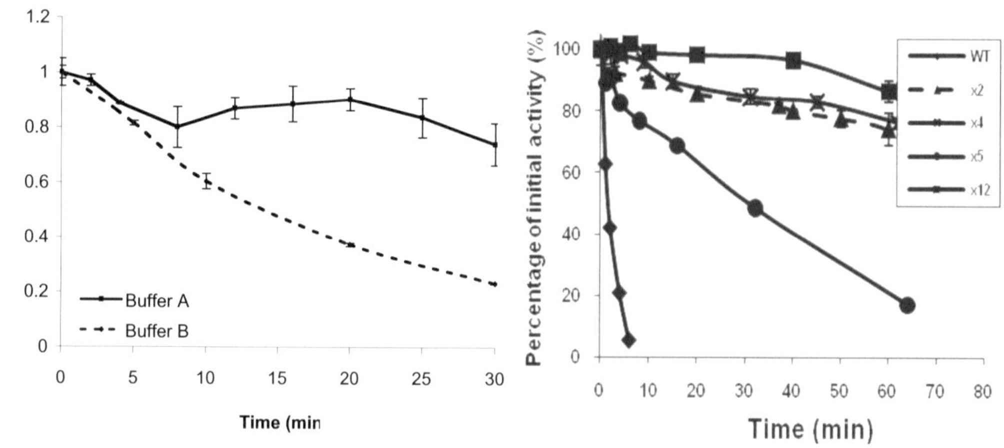

A B

Thermal inactivation of x12 Fluc was determined at 55°C in two different conditions, namely 200 nM of ×12 Fluc in Buffer A (50 mM phosphate buffer, pH 7.8, 10% glycerol (v/v), 2 mM DTT) and 86 nM of ×12 enzyme in Buffer B (20 mM Tris.Cl, pH 8.8, 10 mM KCl, 10 mM $(NH_4)_2SO_4$, 0.1% Triton X-100 (v/v), 5% trehalose (w/v), 0.5% BSA (w/v), 0.4 mg/ml PVP, 10 mM DTT). 30 μl aliquots of enzyme in the respective condition were incubated in water bath at 55°C for varying lengths of time up to 30 min. Enzyme activity was assayed by the injection of 100 μl of TEM, pH 7.8, 1 mM ATP, 200 μM LH_2 into wells containing 5 μl of enzyme and the measurement of flash height. PMT voltages used were 760 mV and 1000 mV for experiment in Buffer A and B respectively. Results shown are mean values ± S.E.M. for triplicate measurements (A). Flash-based activity with LH_2 was compared in aliquots of 0.5 μM enzyme incubated at set temperatures over time. Samples equilibrated to room temperature before dispensing 260 μl 70 μM LH_2 and 1 mM ATP solution in TEM buffer (pH 7.8) into 40 μl luciferase mutant (B).

Fig. 2. Thermal inactivation of x12 Fluc and subset mutants.

3.2 Effect of pH on x12 Fluc activity

Detailed investigation on the pH-dependence of luciferase mutant activity revealed a significant further improvement in pH-tolerant profile from that of ×5 Fluc (Law *et al.*, 2006). The normalised pH-dependence of activity was shown to facilitate comparison of activity across the range of pH values (Fig. 3A). The non-normalised results for ×12 Fluc and UG emphasize the increase in activity for the ×12 Fluc relative to UG (Fig. 3B). The high level of activity (≥ 80 % of maximum activity) exhibited by ×12 Fluc across a range of physiologically relevant pH values (6.6 – 8.6) is likely to offer greater sensitivity and reliability when used in place of existing luciferase mutants in many applications, particularly those requiring a lower pH than the optimal for for Fluc or those that experience pH fluctuations such as whole cell or animal imaging (Frullano *et al.*, 2010).

3.3 Bioluminescence spectra and kinetic properties of WT Fluc, x12 Fluc and UG with LH_2 and ALH_2

The bioluminescence spectrum of WT Fluc is known to undergo a classic red bathochromic shift with LH_2 at low pH, whereas x12 Fluc and UG maintained consistent yellow-green colours emission maximum of 557 nm and 560 nm respectively over the investigated pH range of 6.2-8.8, (Table 2). Neither x12 nor UG showed any significant widening of the spectra at lower pH. Overall, bioluminescence spectra of x12 were even less pH-dependant than those of UG. Both thermostable mutants would be expected to retain their colour under physiological conditions, which may be pertinent to multispectral imaging. The tolerance of x12 Fluc to low pH may give it further advantage in terms of signal strength.

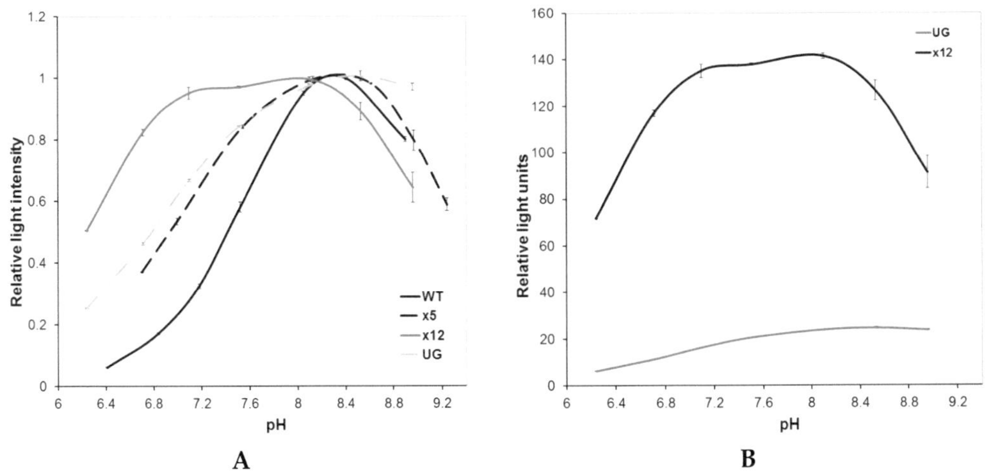

Luciferase activity (20 μl at 0.42 μM) was assayed by the manual mixing with 180 μl of TEM, 1.11 mM ATP, 222 μM LH_2, 300 μM CoA over a range of pH values between 6.0 and 9.5. Bioluminescence was integrated over 5 s using the luminometer at a PMT voltage of 550 mV: normalised to each luciferase enzyme total activity data (A) and non-normalised data (B). Measurements for UG were carried out using a PMT voltage of 700 mV to obtain good signal-to-noise ratio readings; data presented for the non-normalised curve have been corrected for the different PMT voltage used. The lag-time between initiation of the reaction and recording of light emission was ~ 5s. Measurements at each pH were carried out in triplicate. Error bars represent one S.E.M. within triplicate measurements.

Fig. 3. Effect of pH on the activity of WT, UG, x5 and x12 Fluc.

Kinetic parameters including Michaelis-Menten constants (Km) and catalytic constant (Kcat) were measured for the WT, x12 and UG luciferases with LH_2 and ALH_2. The Kms of x12 Fluc (6.8 μM) and UG (3.7 μM) for LH_2 were lower than that of WT Fluc (18 μM) (Table 3), indicating an increase in affinity of the thermostable enzymes for LH_2 relative to the WT. Even stabilising mutations far from the active site can affect tertiary structure and lead to increased substrate affinities and altered enzyme kinetics (Squirrell *et al.*, 1999). Catalytic efficiency characterised by Kcat/Km ratio was significantly compromised in both thermostable luciferases as compared to the WT, with UG being more thermostable but less catalytically efficient than x12. In terms of specific activity x12 Fluc retained only 15% of the specific activity of the WT, exceeding that of UG by four-fold.

For *in vivo* applications involving imaging of protease activity where ALH_2 is to be detected instead of LH_2. Therefore, bioluminescence properties of these mutants with ALH_2 were tested. WT Fluc is known to produce pH independent orange-red emission with ALH_2 (White *et al.*, 1966) and here it was seen that WT Fluc, x12 Fluc and UG were red-shifted to the emission maximum of 601 nm, 592 nm and 585 nm, respectively (Table 4). UG and x12 Fluc also had narrower half-bandwidth with ALH_2 than WT Fluc, which may indicate that more rigid active sites of UG and x12 Fluc confer better protection to the emitter than WT Fluc with ALH_2.

Mutants	Bioluminescent spectra					
	λmax (nm)			Half-bandwidth (nm)		
	pH			pH		
	6.5	7.8	8.8	6.5	7.8	8.8
WT	592	559	558	91	66	66
x12	565	557	558	66	66	67
UG	562	560	559	73	68	70
x2	594	568	568	86	84	85
x4	560	558	561	69	66	70
x5	559	559	558	68	63	62

0.31 nmol of each mutant was added to 150 μM LH_2 and 1 mM ATP in chilled 0.1 M TEM buffer, adjusted to a varying pH and spectra were measured after 30 s at ca. 25°C. Half-bandwidths are widths of spectra at half maximum intensity. Data were corrected for PMT spectral response. UG was assayed at pH 6.2 not 6.5. Standard error ±2nm.

Table 2. Bioluminescent spectra of the WT, Fluc mutants and UG with LH_2 at different pH

Mutants	Michaelis-Menten kinetic parameters			Specific activity	
	Km, μM	Kcat x 10^8, RLU/s	Kcat/ Km x 10^{13}, RLU/ s x μM	x 10^7, RLU/mg	% of WT at pH 7.8
WT	18±1	1129±215	668±157	24.25±0.14	100
x12	6.8±0.2	50±3	74±2	3.56±0.04	15
UG	3.7±0.08	19±2	53±8	0.98±0.02	4
x2	12±2	1260±181	1149±110	11.37±0.03	47
x4	28±3	574±84	201±7	16.59±1.53	68
x5	9.5±1.5	1202±142	1288±51	21.33±0.04	88

Kinetic parameters were derived from flash heights (Hanes, 1932) by varying LH_2 concentration in the range of 0.2 -500 μM in the presence of saturating 1 mM ATP (see Materials and Methods). Specific activity was determined in the presence of 20.1 pmol enzyme by integrating light in 20 ms pulses over 12 s. PMT voltage 550 mV.

Table 3. Kinetic parameters of WT, Fluc mutants and UG with LH_2

In terms of kinetic parameters significant differences were observed in Km and Kcat values for LH_2 and ALH_2 between the WT and thermostable luciferases. In the WT both Km and Kcat for ALH_2 were significantly lower than for LH_2, suggesting that WT Fluc has higher affinity for ALH_2 than for LH_2 (Shinde *et al.*, 2006). In thermostable x12 Fluc both Km and Kcat for ALH_2 were very close to those of LH_2, while in UG Km for ALH_2 was higher and Kcat was lower than for LH_2. The mutations that invoke thermostability increase the affinity of Fluc for LH_2, but reduce it for ALH_2. WT Fluc had nearly 10-fold higher catalytic efficiency (Kcat/Km) with LH_2 and ALH_2 than x12 Fluc. x12 demonstrated comparable Kcat/Km ratio for both substrates, favouring its use with ALH_2 while UG had a similar catalytic efficiency to x12 Fluc for LH_2, but 23-times lower Kcat/Km for ALH_2 . Overall, x12 Fluc seems to be a mutant of choice to be used with ALH_2.

Mutants	Bioluminescent spectra					
	λmax (nm)			Half-bandwidth (nm)		
	pH			pH		
	6.5	7.8	8.8	6.5	7.8	8.8
WT	599	601	601	74	79	79
x12	596	592	591	72	72	71
UG	*588*	585	584	*71*	72	71
x2	596	596	595	76	76	74
x4	596	597	593	74	74	74
x5	599	600	599	74	76	73

0.31 nmol of each mutant was added to 100 μM ALH_2 and 3 mM ATP in chilled 0.1 M TEM buffer, adjusted to a varying pH and spectra were measured after 30 s at ca. 25°C. Half bandwidths are widths of spectra at half maximum intensity. Data were corrected for PMT spectral response. UG was assayed at pH 6.2 not 6.5. Standard error ±2nm.

Table 4. Bioluminescent spectra of the WT, Fluc mutants and UG with ALH_2 at different pH

WT Fluc appears to provide stringent interactions dictating reaction kinetics with ALH_2, but not those that govern the emitting species. This may also be indicated by larger half-bandwidths of bioluminescence spectra with ALH_2 than with LH_2. Low K_m values were proposed to correlate with more blue-shifted emission (Kutuzova *et al.*, 1997), but this is not always true and it is possible for Km to be more linked to bioluminescence spectra half-bandwidth, i.e. the extent of vibrational freedom of oxyluciferin (Viviani *et al.*, 2001). However, here, WT Fluc has wider spectral half-bandwidth with ALH_2 as compared to LH_2, but lower Km. WT Fluc had only 10% of the specific activity with ALH_2 as compared to LH_2 at pH 7.8 (Table 5). On the contrary, x12 Fluc's activity with ALH_2 (6%) is only 2.5-fold lower than that with LH_2 (15%), while UG had as low as 4% with both substrates. Reasons for lower Vm and red-shifted emission of WT Fluc with ALH_2 have been postulated (Shinde *et al.*, 2006; McCapra and Perring, 1985; Wada *et al.*, 2007).

3.4 Properties of thermostable mutants containing subsets of x12 Fluc mutations

Elimination of bathochromic shift at low pH, increase in thermostability and pH-tolerance were associated with the significant reduction in specific activity and changes in the essential kinetic parameters. In an attempt to identify mutations responsible for the undesirable associated changes and to improve the understanding of kinetics and activity differences between WT Fluc and x12 Fluc with LH_2 and ALH_2 a number of thermostable enzymes with subsets of x12 Fluc mutations were investigated (Table 1). These were x2 Fluc (E354R/ D357Y) (Baggett *et al.*, 2004), x4 Fluc (T214A/ I232A/ F295L/ E354K) (Tisi *et al.*, 2002) and x5 Fluc (F14R/ L35Q/ V182K/ I232K/ F465R) (Law *et al.*, 2006). Among the subset mutants only x2 showed bathochromic shift with LH_2 at low pH, similar to that observed in the WT (Table 2). x4 and x5 Flucs resisted red-shift at low pH in line with x12. With ALH_2 none of the subset mutants showed any significant shift in the bioluminescence across the investigated pH range, which was similar to the behaviour of x12, WT and UG with this substrate (Table 4).

Mutants	Michaelis-Menten kinetic parameters			Specific activity	
	Km, μM	Kcat x 10^8, RLU/s	Kcat/ Km x 10^{13}, RLU/ s x μM	x 10^7, RLU/mg	% of WT with LH_2 at pH 7.8
WT	2.4±0.4	174±13	849±158	2.06±0.01	8
x12	6.5±0.2	38±4	59±9	1.46±0.01	6
UG	8.1±2.19	1.4±0.04	2.3±0.7	0.60±0.01	4
x2	2.3+0.1	226±24	956±68	1.99±0.03	8
x4	9±1	144±14	166±14	2.51±0.03	10
x5	1.6±0.1	215±32	1343±184	2.07±0.01	9

Kinetic parameters were derived from flash heights (Hanes, 1932) by varying ALH_2 concentration in the range of 0.1-600 μM used with 3mM ATP to saturate (see Materials and Methods). Specific activity was determined in the presence of 20.1 pmol enzyme by integrating light in 20 ms pulses over 12 s. PMT voltage 550 mV.

Table 5. Kinetic parameters of WT, Fluc mutants and UG with ALH_2

In terms of kinetic parameters, if compared to the WT, the effect of mutations in x4 on Km and Kcat with LH_2 differed significantly from mutations in both x2 and x5 subsets (Table 3). Mutations in x4 increased the Km, while in two other subsets decreased it, and reduced the Kcat, while in the other subsets hardly any change was observed. These results correlated well with the previously reported in literature (Law *et al.*, 2006; Tisi *et al.*, 2002; Prebble *et al.*, 2001; Branchini *et al.*, 2007). Similar trends were observed in the kinetic parameters for ALH_2 (Table 5). Mutations in x4 significantly increased its Km for this substrate (as in x12), while two other mutants slightly decreased it. Kinetics of the x4 mutant with ALH_2 exhibited slower rise and slower decay, reminiscent of the kinetics of x12, whereas x2 and x5 displayed a sharp flash that strongly resembled the WT.

The differences in kinetic parameters and flash kinetics suggested that an undesirable effect of mutations was common to x4 and x12 mutants, but not to x2 or x5, on catalysis with both LH_2 and ALH_2. It was hypothesised that reversion of buried x12 Fluc mutations T214C and F295L might enhance its catalytic properties.

3.5 Properties of T214 and F295 revertant mutants of x12 Fluc

x12 Fluc mutations T214C and F295L were individually reverted and their bioluminescent properties, catalytic parameters and thermal stability investigated. Bioluminescence spectra of both revertants were measured and found to match those of x12 Fluc with both substrates (Table 6). F295 revertant showed no bathochromic shift with either substrate and retain the pH-tolerance previously observed in x12 Fluc. T214 revertant had a flash kinetics similar to x12 Fluc with both substrates, but F295 had a faster and brighter flash with LH_2, more resembling that of the WT (not shown).

Mutants	Substrate	Bioluminescent spectra					
		λmax (nm)			Half-bandwidth (nm)		
		pH			pH		
		6.5	7.8	8.8	6.5	7.8	8.8
x12	LH_2	565	557	558	66	66	67
	ALH_2	596	592	591	72	72	71
T214	LH_2	-	560	-	-	66	-
	ALH_2	-	591	-	-	72	-
F295	LH_2	563	559	560	71	66	70
	ALH_2	593	592	591	70	72	73

Details are as in Table 2 and 4.

Table 6. Bioluminescent spectra of x12 Fluc and its revertants at different pH with LH_2 and ALH_2

T214 revertant exhibited Kms and Kcats similar to x12 Fluc for LH_2 and its Kcat/Km ratio was not significantly different from that of x12 Fluc (Table 7). The Km of T214 for ALH_2 was slightly lower than x12 Fluc and Kcat elevated, resulting in the higher Kcat/Km ratio. T214 revertant had similar specific activities to x12 Fluc with both substrates. However, the Km of F295 revertant for both LH_2 and ALH_2 was lower and Kcat was higher than that of x12 Fluc resulting in a much higher Kcat/Km ratio. Specific activity of F295 revertant with both substrates was two-fold brighter than that of x12 Fluc. As a result of these findings F295 revertant afforded higher sensitivity to the flash-based detection of LH_2 and ALH_2 than x12 Fluc. It is possible to conclude that in x12 Fluc residue F295 contributes much more to catalysis than T214 and its reverse-mutation may significantly improve the performance of x12 Fluc providing it has no negative effect on the thermostability.

Mutants	Substrate	Michaelis-Menten kinetic			Specific activity	
		Km, μM	Kcat x 10^8, RLU/s	Kcat/ Km x 10^{13}, RLU/ s x μM	x 10^7, RLU/mg	% of x12 with LH_2 at pH 7.8
x12	LH_2	6.8±0.2	50±3	74±2	3.56±0.04	100
	ALH_2	6.5±0.2	38±4	59±9	1.46±0.01	40
T214	LH_2	6.2±0.5	49±3	83±10	3.22±0.03	79
	ALH_2	4.9±0.5	66±2	138±8	2.37±0.05	58
F295	LH_2	3.7±0.2	133±6	366±22	8.03±0.09	197
	ALH_2	3.6±0.2	82±2	231±9	3.23±0.02	79

Details are as in Table 3 and 5. Specific activity was measured at pH 7.7 with LH_2 and at pH 8.2 with ALH_2.

Table 7. Kinetic parameters of x12 Fluc revertants with LH_2 and ALH_2

Although T214C and F295L are thermostabilising mutations (Law *et al.*, 2006; Tisi *et al.*, 2002), revertants displayed resistance to thermal inactivation similar to that of x12 Fluc at 40°C and 50°C (Fig. 4). Therefore, the mutation F295L is not essential for practically useful thermostability properties of x12 Fluc, but in the framework of this enzyme contributes to reduced activity and catalytic efficiency with both LH_2 and ALH_2. F295 revertant was hereafter named x11 Fluc. Cumulative addition of multiple phenotype-inducing mutations may enhance desired mutant properties, but some additions can clash and unduly disrupt properties.

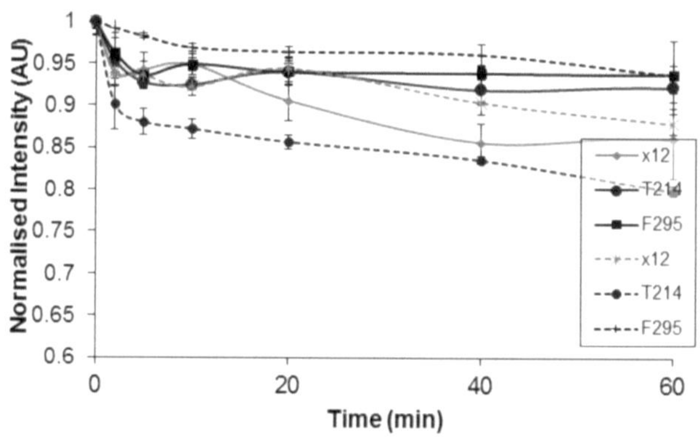

Flash-based activity with LH_2 was compared in aliquots of 0.5 μM enzyme incubated at set temperatures over time. Samples were equilibrated to RT before dispensing 260 μl of 70 μM LH_2 and 1 mM ATP solution in TEM buffer (pH 7.8) onto 40 μl luciferase mutants. Solid lines: 40°C, dashed lines: 50°C.

Fig. 4. Thermal inactivation of x12 Fluc and revertants at 40°C and 50°C.

3.6 *In vivo* imaging

Human retrovirus encoding *Ppy* Fluc or x11 Fluc genes bearing a myc tag was used to transduce Raji cells, which were sorted to the same expression levels (Fig. 5A) by flow cytometric sorting for myc tag staining and cultured at 37°C. As an example of *in vivo* imaging using x11 Fluc, one million Raji cells expressing similar levels of WT or x11 Flucs were injected into the tail veins of immunocompromised Beta2m-mice to induce systemic lymphoma (Chao *et al.*, 2011) and imaged after i.p. administration of LH_2. Images revealed light signals predominantly from brain, spine and hips. x11 Fluc appeared very bright *in vivo* (Fig. 5B) because of its high thermostability and pH tolerance along with favourable kinetic parameters demonstrated in characterisation. It is expected to perform equally well under the changing physiological pH conditions and in combination with aminoluciferin used for imaging of protease assays *in vivo*. Further work to codon-optimise and test this mutant in mammalian cells is underway and will be published shortly.

4. Conclusion

In the present study, we describe the construction and characterisation of the firefly luciferase mutant ×12 Fluc based on the mutations previously identified as increasing thermostability of the enzyme. Detailed characterization of its bioluminescent and biochemical properties revealed that it is the only luciferase mutant reported to exhibit ≥ 80 % of total activity across a wide pH range of 6.6 - 8.8 covering physiologically significant pH at the lower end. Additionally, it possesses sufficient thermal stability to be applicable to assays that require temperatures lower than 50°C or in assays involving short lengths of higher temperature exposure. This mutant could be beneficial for *in vivo* imaging with luciferin, particularly in assays that experience pH fluctuations, and for bioluminescence protease assays or *in vivo* protease imaging, in which the sensitivity of detection may be dependent on the sensitivity of ALH_2 detection. The latter assays could benefit further from increased brightness.

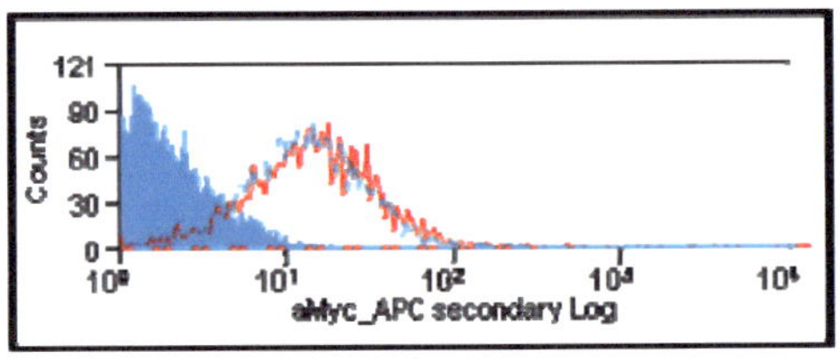

A.

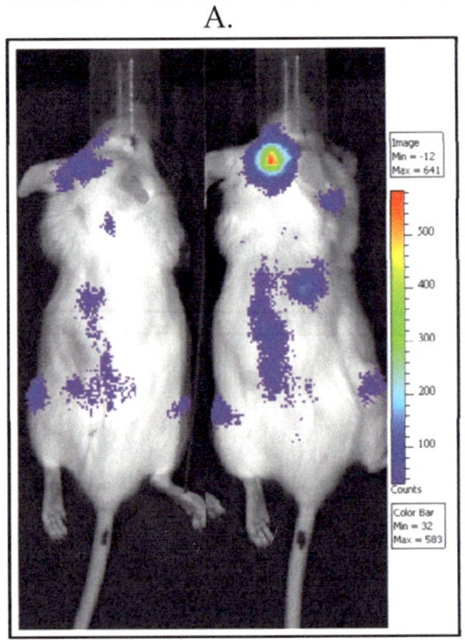

B.

FACS histogram showing expression of WT Fluc and thermostable mutants in Raji cells. Cells obtained by fluorescence activated cell sorting were expanded *in vitro* and expression of transgenes was analysed using a Cyan flow cytometer using anti-myc.FITC staining. Non-transduced cells (blue line – filled) were compared to WT Fluc (red line) and x11 Fluc (blue line – dashed) (A). *In vivo* imaging of systemic lymphoma expressing WT Fluc (left) and x11 mutant (right). Mice with lymphomas expressing either WT Fluc or x11 Fluc were imaged in three groups of three mice (B).

Fig. 5. Expression and example of *in vivo* imaging of Flucs in mammalian cells.

The brightness of x12 Fluc was improved by identifying the mutation at the 295 position as having the major negative impact on the bioluminescent characteristics, reverting it and creating x11 Fluc, which possessed all the desirable properties. x11 Fluc has remarkably high activity and catalytic efficiency with LH_2 and ALH_2, coupled to a high resistance to thermal inactivation and pH-tolerance. Its highly advantageous properties in terms of stability and brightness in mammalian cells have been demonstrated using systemic lymphoma expressing the mutant in mice as an example, as there was no normalisation for engraftment and no statistical difference between WT and x11 in this small sample set.

5. Acknowledgements

Work was funded by a BBSRC CASE award and carried out at the Institute of Biotechnology, University of Cambridge, UK. *In vivo* imaging experiments were carried out at the Cancer Institute, University College London, UK. FACS cell sorting was carried out by Arnold Pizzey (UCL Cancer Institute, UCL, London). Manuscript was proof-read by Nigel Appleton (Lumora Ltd).

6. References

Baggett, B.; Roy R., Momen, S., Morgan, S., Tisi, L., Morse, D. & Gillies, R. (2004). Thermostability of Firefly Luciferases Affects Efficiency of Detection by in vivo Bioluminescence. *Molecular Imaging,* No.3, pp. 324-332

Berger, F.; Paulmurugan, R., Bhaumik, S. & Gambhir, S. (2010). Uptake Kinetics and Biodistribution of ^{14}C-D-Luciferin - a Radiolabeled Substrate for the Firefly Luciferase Catalyzed Bioluminescence Reaction: Impact on Bioluminescence-Based Reporter Gene Imaging. *European Journal of Nuclear Medicine and Molecular Imaging,* No.35, pp. 2275-2285

Bradford, M. (1976). A Rapid and Sensitive Method for the Quantitation of Microgram Quantities of Protein Utilizing the Principle of Protein-Dye Binding. *Analytical Biochemistry,* No.72, p.248

Branchini, B.; Ablamsky, D., Davis, A., Southworth, T., Butler, B., Fan, F., Jathoul, A. & Pule, M. (2007). Red-Emitting Luciferases for Bioluminescence Reporter and Imaging Applications. *Analytical Biochemistry,* No.396, pp. 290-297

Branchini, B.; Ablamsky, D., Murtiashaw, M., Uzasci, L., Fraga, H. & Southworth, T. (2007). Thermostable Red and Green Light-Producing Firefly Luciferase Mutants for Bioluminescent Reporter Applications, *Analytical Biochemistry,* No.361, pp. 253-262

Branchini, B.; Magyar, R., Murtiashaw, M. & Portier, N. (2001). The Role of Active Site Residue Arginine 218 in Firefly Luciferase Bioluminescence. *Biochemistry,* No.40, pp. 2410-2418

Caysa H., Jacob R., Müther N., Branchini B., Messerlec M. and Söling A. (2009). A Red-Shifted Codon-Optimized Firefly Luciferase is a Sensitive Reporter for Bioluminescence Imaging. *Photochemical and Photobiological Sciences,* No.8, pp. 52–56

Chao, M.; Tang, C.; Pachynski, R.; Chin, R.; Majeti, R. & Weissman, I. (2011). Extra-nodal dissemination of non-Hodgkin's lymphoma requires CD47 and is inhibited by anti-CD47 antibody therapy. Blood. Epub ahead of print: DOI: 10.1182/blood-2011-02-338020.

De Luca, M. & McElroy, W. (1974). Kinetics of the Firefly Luciferase Catalysed Reactions. *Biochemistry,* No.13, pp. 921-925

Dragulescu-Andrasi, A.; Liang, G. & Rao, J. (2009). In vivo Bioluminescence Imaging of Furin Activity in Breast Cancer Cells Using Bioluminogenic Substrates. *Bioconjugate Chemistry,* No.20, pp. 1660-1666

Ellis, R. & Wright, A. (1999). Optimal Use of Photomultipliers for Chemiluminescence and Bioluminescence Applications. *Luminescence,* No.14, pp. 11-18

Foucault, M.; Thomas, L., Goussard S., Branchini B. & Grillot-Courvalin, C. (2010). In vivo Bioluminescence Imaging for the Study of Intestinal Colonization by Escherichia coli in Mice. *Applied Environmental Microbiology,* No.76, pp. 264-274

Frullano, L.; Catana, C., Benner, T., Sherry, A. & Caravan, P. (2010). Bimodal MR–PET Agent for Quantitative pH Imaging. *Angewandte Chemie International Edition,* No.49, pp. 2382–2384

Hall, M.; Gruber, M., Hannah, R., Jennens-Clough, M. & Wood, K. (1999). Stabilisation of Firefly Luciferase Using Directed Evolution, In *Bioluminescence and Chemiluminescence: Perspectives for the 21st Century.* Editors: Roda, A.; Pazzagli, M. Kricka, L. Stanley, P. pp. 392-395. John Wiley & Sons Inc., ISBN 9780471987338, 10th I.S.B.C. at Bologna, Italy, Sept. 1998

Hanes, C. (1932). Studies on Plant Amylases: The Effect of Starch Concentration upon the Velocity of Hydrolysis by the Amylase of Germinated Barley. *Biochemical Journal,* No.26, pp. 1406–1421

Kutuzova, G.; Hannah, R. & Wood, K. (1997). Bioluminescence Color Variation & Kinetic Behaviour Relationships Among Beetle Luciferases, In *Bioluminescence and Chemiluminescence: Molecular Reporting with Photons.* Editors: Hastings, J.; Kricka, L. & Stanley P. pp 248-252. John Wiley & Sons Inc., ISBN 9780471975021, 9th I.S.B.C. at Woods Hole, MA., U.S.A., 9th Oct. 1996

Kung A. (2005). Harnessing the Power of Fireflies and Mice for Assessing Cancer Mechanisms. *Drug discovery today: disease mechanisms,* No.2, pp. 153-158

Law, G.; Gandelman, O.; Tisi, L.; Lowe, C. & Murray, J. (2002). Altering the Surface Hydrophobicity of Firefly Luciferase. In *Bioluminescence and Chemiluminescence: Progress and current applications.* Editors: Stanley, P. & Kricka L. pp. 189-92. World Scientific, Singapore

Law, G.; Gandelman, O., Tisi, L., Lowe, C. & Murray, J. (2006). Mutagenesis of Solvent Exposed Amino Acids in Photinus pyralis Luciferase Improves Thermostability and pH-Tolerance. *Biochemical Journal.* No.397, pp. 305-312

Luker K.; Gupta M. & Luker G. (2008). Imaging CXCR4 Signalling with Firefly Luciferase Complementation. *Analytical Chemistry,* No.80, pp. 5565-5573

McCapra, F. & Perring, K. (1985). Luciferin Bioluminescence. In *Chemi-and Bioluminescence.* Editor: Burr J., pp 359-386. CRC Press, 9780824772772, 5th Aug. 1985

McElroy, W. & Green, A. (1956). Function of Adenosine Triphosphate in Activation of Luciferin. *Archives of Biochemistry and Biophysical,* No.64, pp. 257-271

Mezzanotte L.; Fazzina R., Michelini E., Tonelli R., Pession A., Branchini B. & Roda A. (2010). In vivo Bioluminescence Imaging of Murine Xenograft Cancer Models with a Red-Shifted Thermostable Luciferase. *Molecular Imaging and Biology,* No.12, pp. 406-414

Michelini E.; Cevenini L., Mezzanotte L., Ablamsky D., Southworth T., Branchini B. & Roda A. (2008). Combining Intracellular and Secreted Bioluminescent Reporter Proteins for Multicolor Cell-Based Assays. *Photochemical and Photobiological Sciences,* No.7, pp. 212-217

Monsees, T.; Geiger, R. & Miska, W. (1995). A Novel Bioluminogenic Assay for α-chymotrypsin. *Journal of Chemiluminescence and Bioluminescence,* No.10, pp. 213-218

Murray, L.; Luens, K., Tushinski, R., Jin, L., Burton, M., Chen, J., Forestall, S. & Hill, B. (1999). Optimization of Retroviral Gene Transduction of Mobilized Primitive Hematopoietic Progenitors by Using Thrombopoietin Flt3, and Kit Ligands and RetroNectin Culture. *Human Gene Therapy,* No.10, pp. 1743-1752

Nakatsu, T.; Ichiyama, S., Hiratake, J., Saldanha, A., Kobashi, N., Sakata, K. & Kato, H. (2006). Structural Basis for the Spectral Difference in Luciferase Bioluminescence. *Nature Letters,* No.400, pp. 372-376

Prebble, S.; Price, R.L., Lingard, B., Tisi, L. & White, P. (2001). Protein Engineering and Molecular Modelling of Firefly Luciferase. In *Proc. 11th Int. Symp. Biolum. Chemilum.* Editors: Case, J.; Herring, P., Rodinson, B., Haddock, S., Kricka, L. & Stanley, P. pp 181-184. World Scientific Publishing Co, 981024679X, Pacific Grove, CA., U.S.A.

Sandalova, T. & Ugarova, N. (1999). Model of the Active Site of Firefly Luciferase. *Biochemistry (Moscow),* No.64, pp. 1141-1150

Seliger, H. & McElroy, W. (1959). Quantum Yield in the Oxidation of Firefly Luciferin. *Biochemical and Biophysical Research Communications.* No.1, pp. 21-24

Seliger, H. & McElroy, W. (1960). Spectral Emission and Quantum Yield of Firefly Bioluminescence. *Archives of Biochemistry and Biophysics,* No.88, pp. 136-141

Shinde, R.; Perkins, J. & Contag, C. (2006). Luciferin Derivatives for Enhanced in vitro and in vivo Bioluminescence Assays. *Biochemistry,* No.45, pp. 11103-11112

Squirrell, D.; Murphy, M., Price, R., Lowe, C., White, P., Tisi, L. & Murray, J. (1999). Thermostable Photinus pyralis Luciferase Mutant. *U.S. Patent 7906298,* filed October 26, 1999

Tisi, L.; Lowe, C. & Murray, J. (2001). Mutagenesis of Solvent-Exposed Hydrophobic Residues in Firefly Luciferase. In *Proc. 11th Int. Symp. Biolum. Chemilum.* Editors: Case JF, Herring PJ, Robinson BH, Haddock SHD, Kricka LJ and Stanley PE. World Scientific, Singapore. pp. 189-92

Tisi, L.; White, P., Squirrell, D., Murphy, M., Lowe, C. & Murray, J. (2002). Development of a Thermostable Firefly Luciferase. *Analytica Chimica Acta,* No.457, pp. 115-123

Tisi, L.; Law, E., Gandelman, O., Lowe, C. & Murray, J. (2002b). The Basis of the Bathochromic Shift in the Luciferase from Photinus pyralis. *Bioluminescence and Chemiluminescence: Progress and Current Applications.* Editors: Stanley, P. & Kricka, L., pp 57-60. World Scientific Publishing Co., ISBN: 9812381562, Cambridge, U.K., 5th April 2002

Ugarova, N. (1989). Luciferase of Luciola mingrelica Fireflies. Kinetics and Regulation Mechanism. *Journal of Bioluminescence and Chemiluminescence,* No.4, pp. 406-418

Van de Bittner G.; Dubikovskaya, E., Bertozzi, C. & Chang C. (2010). In vivo Imaging of Hydrogen Peroxide Production in a Murine Tumor Model with a Chemoselective Bioluminescent Reporter. *PNAS,* No.14, pp. 21316-21321

Viviani, V.; Uchida, A., Suenaga, N., Ryufuku, M. & Ohmiya, Y. (2001). Thr 226 is a Key Residue for Bioluminescence Spectra Determination in Beetle luciferases. *Biochemical and Biophysical Research Communications,* No.280, pp. 1286-1291

Wada, N.; Fujii, H. & Sakai, H. (2007). A Quantum–Chemical Approach to the Amino Analogs of Firefly Luciferin. *Proc. 14th Int. Symp. Biolum. Chemilum.: Chemistry, biology and applications.* pp 243-246

White, P.J., Leslie, R.L., Lingard, B., Williams, J.R. and Squirrell, D.J. (2002). Novel in vivo reporters based on firefly luciferase, *Bioluminescence and Chemiluminescence: Progress and Current Applications.* Editors: Stanley, P. & Kricka, L., pp 509-12. World Scientific Publishing Co., ISBN: 9812381562, Cambridge, U.K., 5th April 2002

White, P., Squirrell, D., Arnaud, P., Lowe, C. and Murray, J. (1996). Improved Thermostability of the North American Firefly Luciferase: Saturation Mutagenesis at Position 354. *Biochemical Journal,* No.319, pp. 343-350

White, E.; Worther, H., Seliger, H. & McElroy, W. (1966). Amino Analogs of Firefly Luciferin and Activity Thereof. *JACS,* No. 88, pp. 2015-2018

Willey, T.L., Squirrell, D. and White, P. (2001). Design and Selection of Firefly Luciferases with Novel in vivo and in vitro Properties. In *Proc. 11th Int. Symp. Biolum. Chemilum.* Case, J.; Herring, P., Robinson, B., Haddock, S., Kricka, L. & Stanley, P. pp. 201-204, World Scientific, Singapore. ISBN: 981024679X

Wood K. (1998). The Chemistry of Bioluminescent Reporter Assays. *Promega Notes,* No.65, pp. 14-21

Woodroofe, C.; Shultz, J., Wood, M., Cali, J., Daily, W., Meisenheimer, P. & Laubert, D. (2008). N-Alkylated 6′-Aminoluciferins are Bioluminescent Substrates for Ultra-Glo and Quantilum Luciferase: New Potential Scaffolds for Bioluminescent Assays. *Biochemistry,* No. 47, 10383-10393

Zinn K.; Chaudhuri, T., Szafran, A., O'Quinn, D., Weaver, C., Dugger, K., Lamar, D., Kesterson, R., Wang, X. & Frank, S. (2008). Noninvasive Bioluminescence Imaging in Small Animals. *ILAR J,* No.49, pp. 103-115

7

Bioluminescence Applications in Preclinical Oncology Research

Jessica Kalra[1,2] and Marcel B. Bally[1,3,4,5]
[1]*Experimental Therapeutics BC Cancer Agency,*
[2]*Langara College, Vancouver, BC,*
[3]*Department of Pathology and Laboratory Medicine,*
University of British Columbia, Vancouver, BC,
[4]*Faculty of Pharmaceutical Sciences, University of British Columbia, Vancouver, BC,*
[5]*Centre for Drug Research and Development, Vancouver, BC,*
Canada

1. Introduction

In vitro studies have offered vast insight into much of cancer biology, however, it is widely accepted that cell based assays are unable to provide a complete picture when attempting to understand the dynamic nature of cancer as it behaves in situ. Critical processes to cancer progression such as angiogenesis, metastasis and response to treatment, rely on complex interactions between tumor cells and their microenvironment. To overcome this challenge, xenografts have been widely used to study cancer biology within the context of a whole organism since as early as the 1950's. These models rely on use of murine (syngeneic) and human (allogeneic) tumor cell lines injected subcutaneously into a rodent host. The subcutaneous animal model has been a valuable tool in the study of cancer and has directly led to the validation of many of the anticancer agents which benefit patients today. Subcutaneous tumor models are easy to implement and monitor due to the accessibility of the tumor tissue. Evaluation of subcutaneous tumors involves calliper measurements of tumor size (width, length and/or height), which are then used to define tumor volume. However, like cell-based assays, subcutaneous tumor models have proven to be poor predictors of therapeutic activity in patients and this is likely due to the reliance on cell lines which when inoculated subcutaneously develop tumors that poorly mimic the biological behaviour of human disease. Cancers arise slowly and evolve into a heterogeneous structure both in terms of cellular composition (host cells and tumor cells) and microenvironment (vascularization and transient regions of hypoxia and nutrient stress). Subcutaneous xenografts are implanted in microenvironments that will be remarkably different from the tissue of origin. This means subcutaneous tumor cells do not receive the same signals from the stroma that influence immunity, angiogenesis and metastasis; all factors that impact tumor progression and response to therapeutic interventions. Although the initiators and drivers of tumors in humans remain poorly understood, it is generally accepted that following initiation, endogenous disease progresses into a primary tumor which in time can invade surrounding tissues. The latter process involves both extravasation and intravasation of cancer cells. Thus if left untreated primary malignancies can evolve into a metastatic disease which ultimately engender systemic changes that are incompatible with life. Even this very simplistic description of cancer biology highlights the serious shortcoming of subcutaneous tumor models derived following injection of cultured tumor cell lines. The limitations are even more profound when considering changes in tumor biology that occur as a consequence of treatments.

In recent years orthotopic inoculation of tumor cells has be viewed as a reasonable alternative for initiation of model tumors. In these models, tumor cells are injected in a site that represents the tissue of origin and thus may be closer in characteristics to the original microenvironment. Orthotopic models can exhibit tumor growth rates, capacity for angiogenesis and an invasive potential that better mimic the evolution of cancers in situ. In several examples primary tumors arising following orthotopic injection of tumor cells have been shown to metastasize through lymphatic drainage and/or hemotological spread in ways that are comparable to that seen in human disease. For this reason, some investigators believe that orthotopic tumors more accurately reflect human disease, and may serve to better predict therapeutic outcomes.

Further, in an effort to model systemic disease a variety of cell inoculation methods can be employed to promote haematological spread of tumor cells. These methods include, but are not limited to, tail vein and intracardiac injections. Many groups have used different inoculation methods to assess specific metastatic sites, such as intratibial inoculations of prostate cancer cells in order to study bone metastases, or intraperitoneal injections of ovarian cancer cells to study disease development/progression in the peritoneal cavity. Finally, transgenic animal models are now readily available, where oncogenes and tumor suppressor genes relevant to a specific cancer are knocked in or knocked out. These genetically engineered animals have been useful for studies exploring how genetic alterations are linked to carcinogenesis.

A very large obstacle in studies using orthotopic, systemic and transgenic animal models is monitoring disease burden. The tumour tissue is often inaccessible for visual inspection, localized in organs deep within the body. Assessment of such tumors requires termination of animals at various time-points following disease initiation or at a time when the animal experiences signs of distress/illness. Mice are then euthanized and organs removed for gross and histological assessment of primary and metastatic disease; a practice that frequently requires serial sacrifice and large numbers of animals for a single study. Additionally, in this type of study design, comparisons are made between different groups of animals that were sacrificed at different time-points. Due to animal to animal variations, comparisons are often difficult to interpret and conclusions may be over- or even understated.

The use of orthotopic and transgenic cancer models has fostered development of small-animal imaging methods to follow tumour development and progression in live animals. There are several modalities that are applicable to small animal imaging, including ultrasound (US), magnetic resonance imaging (MRI), computed tomography (CT), positron emission tomography (PET), and single photon emission computed tomography (SPECT). Each of these imaging modalities have strengths and weaknesses as recently reviewed by Ray et al (Ray 2011). Over the last decade use of Bioluminescence Imaging (BLI) has become increasingly popular in part because of the accessibility of imaging tools and because the method is extremely sensitive with a capability of detecting as few as 10 tumor cells in a live animal. BLI provides a non-invasive, semi-quantitative approach to localizing small tumors, following growth and metastasis and monitoring tumor response to treatment in the same animal longitudinally. This non-invasive determination of tumor burden over time reduces the numbers of animals required for experiments and provides information on the various stages of tumor development in the same animal as the disease progresses.

The first experiments using BLI for monitoring tumor phenotypes and response to therapy were performed by assessing luciferase activity as a measure of metabolism as described in section 6.0. Recently, more complex studies have been designed. For example, our group used BLI to track the development of an experimental metastatic model of breast cancer after an intracardiac injection of tumor cells. Further, we evaluated the use of an existing and clinically relevant drug to treat orthotopic, metastatic and ascities disease and correlated changes in tumor burden as measured by BLI to pharmacokinetic data in the same animals. The utility of BLI in assessing drug efficacy is multi-faceted in that it is able to address semi-quantitatively the issue of disease burden, and also to assess disease physiology. For example, as firefly luciferase (F-Luc) activity is dependent on the presence of oxygen and ATP, photons are only emitted from metabolically active cells. Thus, therapeutic effects involving changes in tumor metabolism can be readily assessed, where necrotic regions within a tumor can be identified and potentially act as a marker for a positive drug response. The high sensitivity of BLI also allows for the detection of small numbers of tumor cells very early in the development of primary or metastatic disease; cancer cells can be visualized using BLI before they can be visualized by other imaging methods. Over a very short period of time, studies involving BLI have demonstrated that the technique is highly sensitive, high throughput, and relatively easy to use. It is likely that the use of BLI over the next decade will continue to increase in its complexity and its elegance.

In this chapter, the information gleaned from oncology research using BLI is summarized. The strengths, weaknesses and major findings from the last twenty years are consider in an attempt to exemplify the utility of BLI as well as illustrate some of its limitations. Major problems associated with this imaging modality are recognized in order to assist in designing preclinical experiments for those using this imaging modality. Topics summarized below include the development of luciferase positive orthotopic, metastatic and genetically engineered models of human cancer as well as the use of BLI for the assessment of therapeutic activity of drug candidates, as a tool for monitoring gene delivery and gene expression in vivo, for assessment of processes such as angiogenesis and apoptosis, and, finally, for imaging of metastasis and minimal disease in cancer models.

2. Bioluminescence

Bioluminescence (BL) is defined as the production of light by a living organism. Many organisms such as bacteria, fungi, fish, marine invertebrates, and insects use BL for the purpose of mating, camouflage, repulsion, communication and illumination. The chemical reaction that produces BL requires a pigment known generally as luciferin and enzymes called luciferase (see Reaction 1). The reaction between luciferase and its substrate is an oxidation reaction which is sometimes mediated by cofactors such as calcium and may require energy in the form of ATP.

Reaction 1 Luciferin + $O_2 \rightarrow$ oxyluciferin + light

Luciferins are a family of light emitting proteins that act as substrates to luciferase. Luciferins evolved many times in various organisms, hence there are a variety of luciferins that are structurally and chemically distinct (Shimomura 2006). Table 1 provides a list of

Organism	Luciferin structure	Co-factors	Luminescece
Firefly (Photinus pyralis) Clickbeetle (Pyrophorus plagiophthalamus)	D-Luciferin 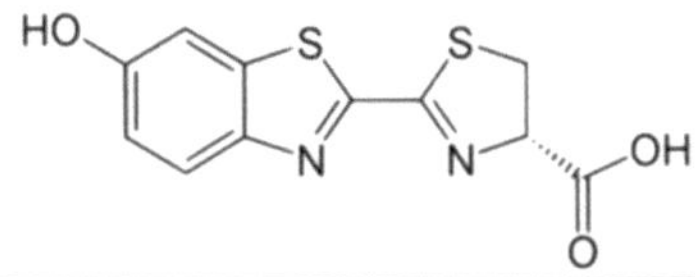	ATP Mg2+	560-615nm
Sea Pansy (Renilla reniformis) Marine copepod (Gaussia princeps)	coelenterazine 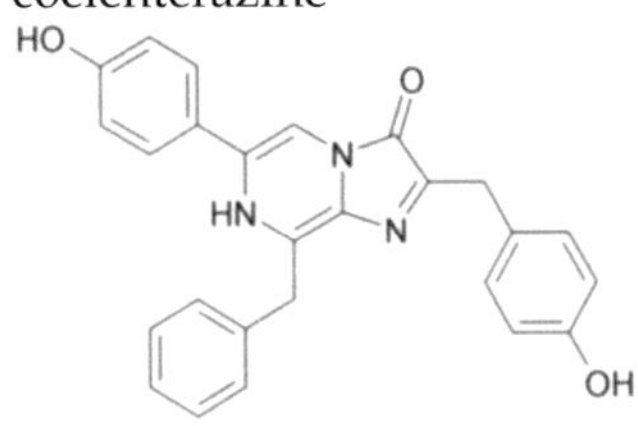	Ca2+	480nm
Snail (Latia neritoides)	Latia 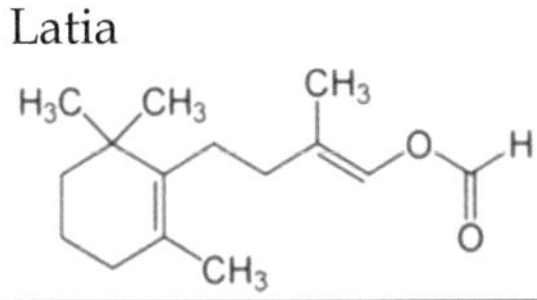	O2	536
Marine Ostracod (Vargula Hilgendorfi)	Cypridina 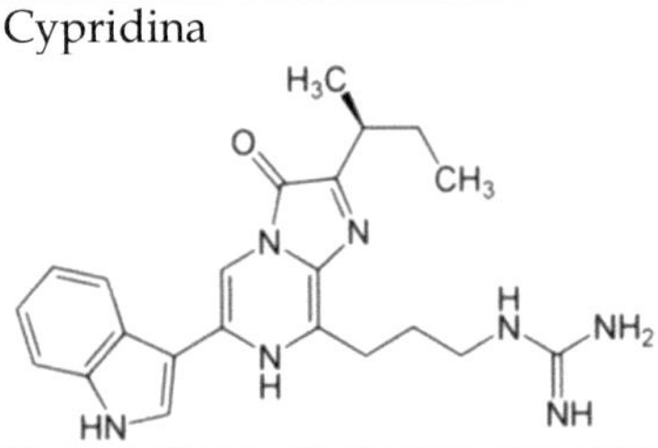	O2	450-460nm
Dinoflagellates	Tetrapyrrole 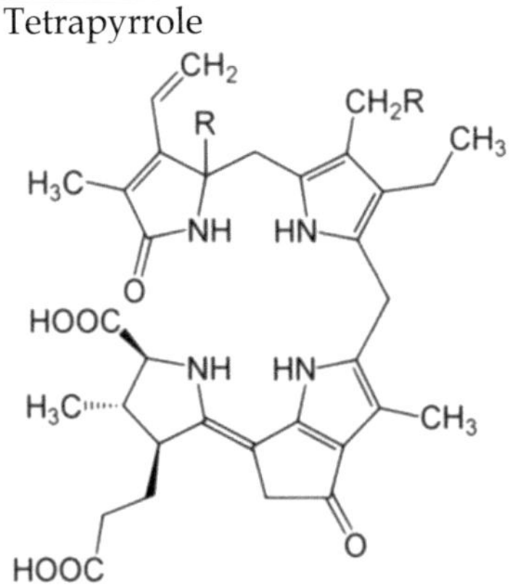	O2	474nm

Table 1. The structure, luminescence wavelength and cofactors required by the luciferases and luciferins used in biological and chemical research (Shimomura 2006)

luciferins that are currently used in biological research. Luciferase, is the general term for a class of enzymes which catalzye oxidation reactions involving luciferins. The structure of luciferase seems to have an impact on the wavelength of photons emitted and color of light produced so that reactions may exhibit yellow-green to red light. Although many of the luciferase enzymes discussed in Table 1 have been applied to in vitro and in vivo work, firefly luciferase (F-Luc) is most commonly employed in oncology research. In the chemical reaction between F-Luc and its substrate D-luciferin, ATP and oxygen promote the formation of oxy-luciferin species. The emission spectrum of the F-Luc catalyzed reaction is broad with an emission peak at approximately 560 nm and a large component above 600 nm. Yellow-green to yellow-orange light is emitted following the relaxation of oxy-luciferin to its ground state (see reaction 2).

Reaction 2

D-Luciferin + ATP → (reversible uses luciferase and Mg2+) luciferase luciferyl-AMP + PPi

Luciferyl-AMP + oxygen → (uses luciferase and Mg2+) luciferase + oxyluciferin + AMP + CO2 + hv (de Wet, Wood et al. 1987)

3. Constructing bioluminescent mammalian cells and animal models

In the following sections the application of biological models expressing luciferase are described in detail. Mammalian systems do not naturally express luciferase and therefore these systems must be engineered to expresse the enzyme. A vector capable of constitutive expression of luciferase or vectors designed to achieve controlled expression of the enzyme can be used Fusion genes can be used as reporter constructs to track gene delivery and integration. Expression of fusion proteins can be used to follow protein localization and protein-protein interactions. Cells with stable expression of luciferase can be used for in vitro assays or be injected into animals to examine metabolism, immunity, angiogenesis, or to establish disease that can be monitored using BLI. Additionally, transgenic animals may also be developed where reporter gene expression is introduced through the germline.

3.1 Engineering Luciferse expressing cells

Luciferase positive tumor cells used for generating animal models in cancer research are widely available in almost all histological types such as breast, cervical, colorectal, lung, prostate, ovarian cancer and melanoma. Developing a luciferase expressing mammary cell line can be accomplished in house by standard transfection or transduction methodology using reagents, such as plasmid vectors carrying the luciferase genes, which are commercially available.

Tang et al use a typical transfection procedure to create a luciferase positive neuroprogenitor cell line where the pGL3Basic plasmid (Promega, Madison, WI) carrying F-Luc was digested with HindIII and BamHI and the 1.9-kb cDNA fragment encoding F-Luc was isolated and cloned into a second vector (the pHGCX). The resulting pHGCX vector contained the F-Luc gene driven by the cytomegalovirus (CMV) promoter which enables constitutive expression of luciferase. The pHGCX vector also contained the gene for enhanced green fluorescent protein (eGFP) under the control of the viral immediate early

IE4/5 promoter, providing a method of fluorescence selection. The neuroprogeneitor celline (C17.2) were stably cotransfected with the engineered vector and pBabePuro, containing the gene for puromycin resistance, using Lipofectamine transfection reagent (Invitrogen Life Technologies, Carlsbad, CA) (Tang, Shah et al. 2003).

Stable luciferase expressing cells can also be generated through transduction strategies (Nyati, Symon et al. 2002; Kalra, Warburton et al. 2009; Kalra, Anantha et al. 2011; Yan, Xiao et al. 2011). Our own lab uses transduction procedures to co-express F-Luc and GFP in breast cancer cell lines. Briefly, the luciferase coding sequence was isolated from the pGL3Basic vector (Promega, Madison, WI) and cloned into the lentiviral vector, FG9, downstream of the CMV and UBiC promoters again to support constitutive activation of the F-Luc gene. The engineered vector was cotransfected with packaging constructs pRSVREV, pMDLg/pRRE and the VSV-G expression plasmid pHCMVG into a packaging cell line (HEK-293T) by a standard LipofectAMINE 2000 (Invitrogen, Burlington, ON Canada) transfection procedure. Conditioned medium containing Lentivirus-Luciferase (Lenti-Luc) particles was collected and cleared of debris by low speed centrifugation. A similar method was used to generate GFP-expressing lentivirus (Lenti-GFP). The Breast cancer cell line (MDA MB 435/LCC6 (LCC6)) was then infected with Lenti-Luc and Lenti-GFP creating LCC6-Luc/GFP cells. To enrich for luciferase positive cells, cells were sorted by FACS for GFP expression. Subsequently GFP-positive cells were re-plated in low concentrations into soft agar in the wells of a 96-well plate. Luciferin was added to each well and plates were imaged using IVIS to identify luciferase positive colonies (Kalra, Warburton et al. 2009; Kalra, Anantha et al. 2011).

3.2 Use of Luciferase positive cell lines

Once stable luciferase positive cell lines are generated, the cells should be assessed for growth rates and sensitivity to selected drugs; comparing engineered cells to parental cell lines. These assays are performed to ensure that the luciferase gene does not interfere with cell function. Many groups will also image a serial dilution of cells to associate BL produced with number of cells as shown in Figure 1. Figure 1A exhibits the CCD camera capture of a serial dilution of luciferase positive human breast cancer (LCC6-Luc) after luciferin substrate is added. BL data was quantified as photons emitted per second. The resulting graph (Figure 1B) shows that the photons emitted are proportional to the number of cells plated (Kalra, Anantha et al. 2011). These data can be used to estimate cell numbers from BL captured in an in vivo model. In this example the minimum number of cells detectable was approximately 10,000 cells, but the detection limit is dependent on multiple factors including lucifern concentration, imaging programs, parameters such as exposure time and the device used for image capture.

Once luciferase positive cells are adequately vetted in vitro, these cells can be used to establish an in vivo model of disease through multiple routes of inoculation as discussed in Section 6.

3.3 Luciferase expressing transgeneic mice

A wide variety of transgenic reporter animals using luciferase-based technology have been engineered in order to visualize transgene expression in vivo. For example, in 2005 Hsieh et al reported the development of a luciferase transgenic mouse model where F-Luc was placed under the control of the Prostate Specific Antigen (PSA) promoter to develop transgenic PSA-Luc mice. Figure 2A shows the BLI of representative PSA-Luc transgenic male and female animals, and non-transgenic littermate male mice. Figure 2B shows BLI of excised organs from a 12-week-old PSA-Luc male mouse. Hsieh et al demonstrate that the PSA-Luc mouse model has luciferase based BL restricted to the prostate gland (Hsieh, Xie et al. 2005). This mouse model has been used to monitor the prostate gland during development, tumorogenesis and in response to androgens or chemotherapy.

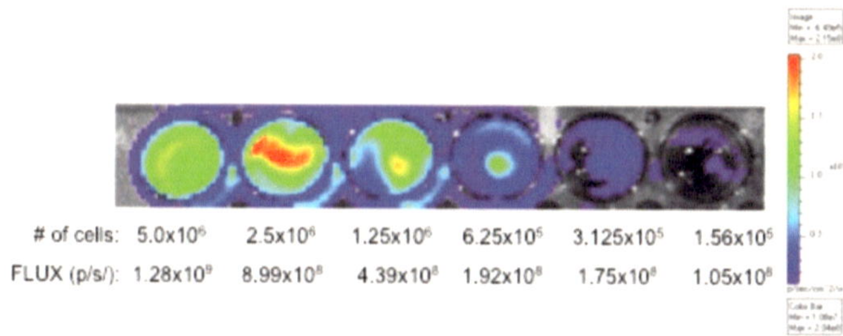

A

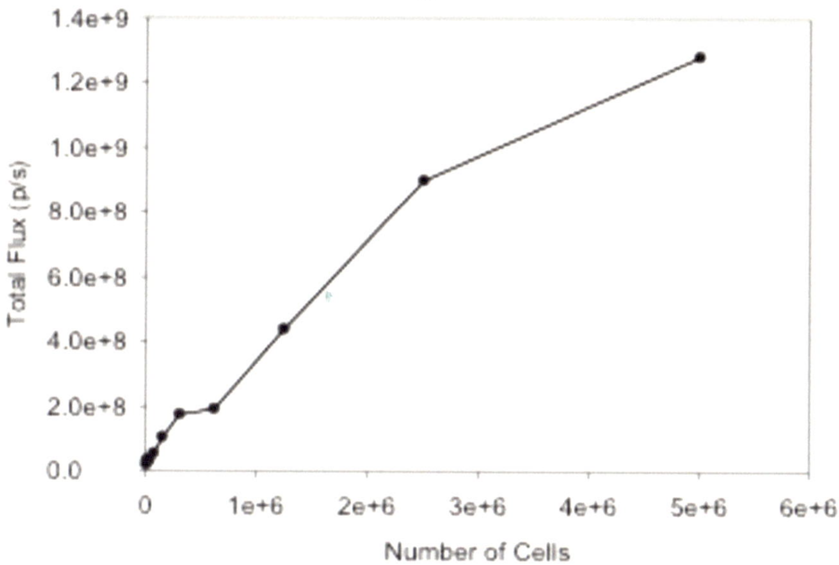

B

Fig. 1. Imaging luciferase positive breast cancer cells in vitro. LCC6-Luc cells were serially diluted and placed into wells of a 24 well plate. Luciferin was added and cells were immediately imaged using the IVIS 200 system to obtain BL measurements (A-representative images). These data were used to generate a plot comparing total light emission to cell number (B - graph). (Kalra, J., M. Anantha, et al. (2011). "Validating the use of a luciferase labeled breast cancer cell line, MDA435LCC6, as a means to monitor tumor progression and to assess the therapeutic activity of an established anticancer drug, docetaxel (Dt) alone or in combination with the ILK inhibitor, QLT0267." Cancer Biol Ther 11(9): 826-838. Reproduced by permission of author.)

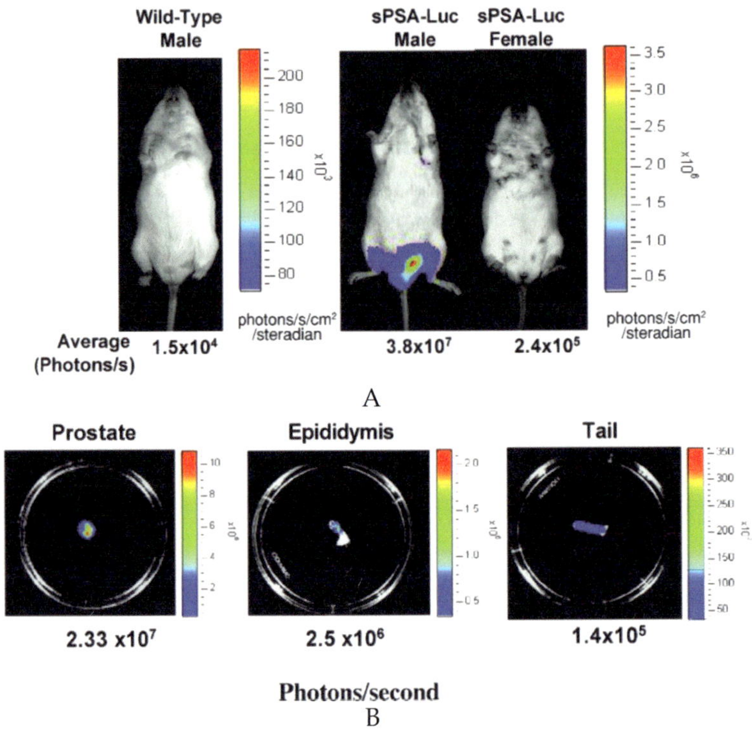

Fig. 2. Bioluminescence imaging of luciferase activity in PSA-Luc transgenic mice. BLI following injection of luciferin in control (A), PSA-Luc transgenic males (B), and PSA-Luc transgenic female (C) mice at 10–12 weeks of age. Isolated prostate, epididymis and tails from 12-week-old PSA-Luc male mice were imaged following luciferin administration (D). (Hsieh, C. L., Z. Xie, et al. (2005). "A luciferase transgenic mouse model: visualization of prostate development and its androgen responsiveness in live animals." J Mol Endocrinol 35(2): 293-304. Reproduced by permission.)

The PSA-Luc animal is just one example of a transgenic mouse models available for BLI. Table 2 summarizes a list of luciferase transgenic mouse models that have been established over the last 10 years and used recently in preclinical studies of cancer. Many of these strains are now commercially available. Goldman et al use the Jackson Laboratory ODD-Luc transgenic mouse strain to construct a spontaneous tumor model with a built in reporter that can be used to localize tumors. The ODD-Luc transgenic mouse expresses HIF-1α oxygen-dependent degradation domain (ODD) fused to luciferase in all tissues, however, under hypoxic stress the ODD-Luc accumulates in hypoxic tissue and is readily observed by BLI. In the Goldman study, the ODD-Luc mice were crossed with Mouse Mammary Tumor Virus (MMTV) transgenic mice. MMTV can act as an insertional mutagen or induce transcription of nearby oncogenes post-insertion leading to the development malignant tumors in the mammary gland of infected mice. The MMTV transgenic mouse has been engineered to express the MMTV-LTR which predisposes the animal to develop multiple spontaneous mammary tumors (Taneja, Frazier et al. 2009). Since solid tumors often contain hypoxic centers, this group demonstrated that the ODD-Luc/MMTV transgenic mouse can be used to follow spontaneous tumor development, progression and response to treatment.

They were able to use BL to image the growth and development of spontaneous tumors and showed that in response to treatment using doxorubicin and prednisone, these tumors were able to regress.

Transgene	Research focus	Reference
Caspase cleavage sequence (ER-DEVD-Luc)	Apoptosis	Laxman 2002 (Laxman, Hall et al. 2002)
p53-Luc	Cell cycle, apoptosis	Rehemtulla 2004(Rehemtulla, Taneja et al. 2004), Briat 2008(Briat and Vassaux 2008)
Smad-responsive element (SBE-Luc)	TGFbeta/Smad signaling	Lin 2005(Lin, Luo et al. 2005), Luo 2009(Luo and Wyss-Coray 2009)
E2F1-Luc	Cell proliferation	Momota 2005(Momota and Holland 2005)
Inducible Nitric Oxide Synthase (FVB/N-Tg(iNOS-Luc)Xen)	Inflammation	Moriyama 2005(Moriyama, Moriyama et al. 2005)
Estrogen Receptor (ER-Luc)	Hormone receptor activity	Wu 2008(Wu, Xu et al. 2008)
Prostate Specific Antigen (PSA-Luc)	Hormone receptor activity	Hiseh 2005(Hsieh, Xie et al. 2005), Lyons 2006(Lyons, Lim et al. 2006), Iyer 2005(Iyer, Salazar et al. 2004; Iyer, Salazar et al. 2005)
Vascular Endothelial Growth Factor (VEGF-Luc)	Angiogenesis	Faley 2007(Faley, Takahashi et al. 2007)
NFkB-Luc	Signaling and inflammation	Vykhovanets 2008(Vykhovanets, Shukla et al. 2008), Robbins 2011(Robbins and Zhao 2011)
EL1-Luc/TAg	Spontaneous pancreatic tumors	Zhang 2009(Zhang, Lyons et al. 2009)
Cre/Lox Luc	Conditional glioblastoma multiforme model	Woolfenden 2009(Woolfenden, Zhu et al. 2009)
Survivin (Survivin –Luc)	Apoptosis	Li 2010(Li, Cheng et al. 2010)
RipTag-IRES-Luc	Pancreatic beta cell carcinogenesis	Zumsteg 2010(Zumsteg, Strittmatter et al. 2010)
Collagen1 alpha 1 (Col-Luc)	Bone metastasis	Lee 2010(Lee, Huang et al. 2010)
Mouse Period 1 (mPer1-luc)	Modulators of circadian rhythm and stromal signaling to tumors	Geusz 2010(Geusz, Blakely et al. 2010)
X-box binding protein 1 (XBP1-Luc)	A marker for stromal stress and reporter for endoplasmic reticulum	Spiotto 2010(Spiotto, Banh et al. 2010)

Transgene	Research focus	Reference
A triple transgenic strain (MMTV-Cre, CAG-betageo-tTA, TetO-Luc)	Spontaneous mammary tumors	Zhang 2010(Zhang, Triplett et al. 2010)
A double transgenic strain (p38DN/ AP-1-Luc)	Cell Cycle and apoptosis	Dickinson 2011(Dickson, Hamner et al. 2007)
HIF-1α oxygen-dependent degradation domain (ODD-Luc)	Hypoxic stress	Goldman 2011(Goldman, Chen et al. 2011)
Human Reverse Telomerase Transcriptase (hTERT-Luc)	Telomerase activity	Jia 2011(Jia, Wang et al. 2011)
Vascular Endothelial Growth Factor Receptor (VEGFR2-Luc or VEGFR2-Luc)	Angiogenesis	Angst 2011(Angst, Chen et al. 2010)
Matrix metalloproteinase 9 (MMP-9-Luc)	Invasion and metastasis	Biron-Pains 2011(Biron-Pain and St-Pierre 2011)
Early growth response 1 (Egr-1-Luc)	Growth factor signaling	Dussman 2011(Dussmann, Pagel et al. 2011)

Table 2. Luciferase based transgenic mouse models with applications in preclinical cancer research

4. Bioluminescent imaging (BLI) and sensitivity

As outlined in the previous section, cells and animal models can be engineered to express luciferase in a variety of ways. However, as implied in the examples above these cells must be exposed to a substrate in order for BL to be produced. Thus although BLI is non-invasive, in vivo luminescence is generated only following intraperitoneal (IP), subcutaneous (SQ), intratumoral (IT), oral (PO) or intravenous (IV) injections of the substrate luciferin. Following injection, up to 15 minutes are required for sufficient distribution of the substrate to sites where luciferase-expressing cells are located and to achieve optimal signal intensity prior to imaging. This step represents the most significant limitation of BLI in models of cancer.

It should also be noted that BL is a weak phenomenon producing low-intensity light that cannot be observed using conventional cameras. Therefore specially designed low light imaging cameras are required. Several commercially available systems are capable of detecting such low levels of light and are listed in Table 3. In general, the components of each system include a light-tight imaging chamber and a super-cooled charged coupled device (CCD) camera. Detected photons emitted from within the body of the animal are converted to a pseudo-colour image representing light intensity (from blue for least intense to red for most intense) and superimposed over the grayscale digital image as shown in Figure 3A. The spatial resolution of BLI, however, is relatively low (1–2 mm), when compared to CT, PET and SPECT. It has been suggested that the poor resolution of BLI is the result of scattering and diffraction of light due to changes in the refractive index at cell membranes and organelles. However, because BLI is associated with little to no background noise, the anatomical origin of photons can be determined, to approximately 1 mm (Edinger,

Cao et al. 2002). Using topographical scanning it is now possible to construct a three dimensional image of the animal at the same time BL data is being collected. This combined imaging may help to provide better resolution and signal localization. The IVIS 200 system is able to create a 3 dimensional image of the animal where a scanning laser positioned in the horizontal plane is used to make a measurement of surface topography as shown in Figure 3B. This image is converted into a digital reconstruction of the animal that can be superimposed onto an animal atlas and used to localize the depth of signal as seen in Figure 3C. Newer BL imaging systems such as the Spectrum CT by Caliper are able to simultaneously create a Computed Tomography scan for the purpose of constructing a three dimensional image of the animal (Figure 3D). Multi-modal imaging can aid in localization, and an assessment of the depth of signal within body cavities with more precision than BLI alone.

Company	System	Method used for supercooling
Caliper LifeSciences	IVIS	Thermoelectric cooling
Andor	iXon	Thermoelectric cooling
Berthold	NightOWL	Pelitier cooling
Hamamatsu +	VIM camera	Intensified cooling
Improvision	Model C2400-47	Cryogenic cooling
Roper Scientific	ChemiPro	Cryogenic cooling
Biospace	PhotoImager	Intensified cooling
Koday	In vivo FX	Thermoelectric cooling

Table 3. Commercially available BL imaging systems (Baert 2008)

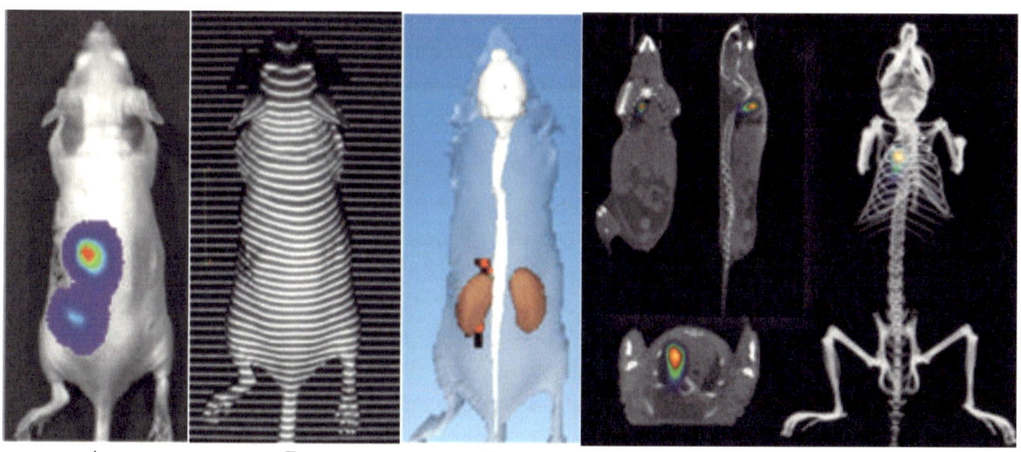

A B C D

Fig. 3. 2D and 3D detection of F-Luc labeled cells using the IVIS system from Caliper Life Sciences. A grayscale digital image is taken at the same time as photon capture, subsequently BL is superimposed onto the digital image (A). Surface topography is constructed using horizontal laser scanning (B), and used to create 3D rendering of the animal which can be overlain on an organ atlas (C). Figure C shows red dots where photon intensity is highest, indicating the depth and localization of the signal. Newer systems incorporate a CT scan with BL (D) in order to gain higher precision in signal localization. (Reproduced by permission from Caliper Life Sciences, Hopkinton, MA, USA.)

Although some of the limitations of this imaging method are summarized above, these limitations are offset by the fact that BLI can detect low levels of gene expression as well as small numbers of cells in animal tumor models. This method can capture minimal disease and/or micro-metastases at distant sites before the appearance of palpable nodules or clinical symptoms. Indeed, signals can be readily visualized immediately following inoculation of tumor cells (Kalra, Anantha et al. 2011). To illustrate the sensitivity of BLI, Lipshutz et al used adenoviral vector carrying the luciferase gene to deliver luciferase to day 15 fetal mice through intraperitoneal delivery. This group was able to image a single luciferase positive liver cell in one million using BLI (Lipshutz, Titre et al. 2003). Sweeney et al were able to use BLI to detect as few as 100-2500 luciferase positive cervical cancer cells in the peritoneal cavity (Sweeney, Mailander et al. 1999). It is also worth noting that in addition to excellent sensitivity, BLI is amenable to medium throughput screening where 6 animals can be imaged at once. Once optimized, the image acquisition times can be less than 30 seconds, and the image acquisition and follow-up analysis are user-friendly.

There are additional advantages associated with BLI in comparison to other imaging methods. The equipment and reagent costs tend to be relatively low compared to PET and/or MRI. BLI is also a non-radioactive imaging modality, in contrast to other modalities such as PET and SPECT. Furthermore, BL light is emitted directly by the specimen without the need to add excitation light, which is required fluorescence imaging. Also photobleaching and phototoxic effects are not a concern. Mammalian tissues and chemical agents such as chemotherapeutics do not normally emit BL, but can generate autofluorescence that can interfere with fluorescent based imaging methods. BLI has low background and any photon emission would be the result of engineered BL cells or genes. Finally, luciferase and its substrate, luciferin, are not toxic to mammalian cells, and no functional differences have been reported between cells expressing luciferase when compared with parental cell lines (Sweeney, Mailander et al. 1999; Edinger, Cao et al. 2002; Tiffen, Bailey et al. 2010; Kalra, Anantha et al. 2011).

When designing animal studies that employ BLI, it is important to consider some factors that can influence the detection of BL photons; factors that make this approach to imaging semi-quantitative at best. These factors include 1) the distance signals must travel through tissues, 2) the nature of overlying structures and 3) cell physiology. For example, luciferase positive cells located deep within the body will appear less bright than an equivalent number of cells located near the surface of the skin (El-Deiry, Sigman et al. 2006; O'Neill, Lyons et al. 2010); as tissues overlying the target cells can attenuate photon emission. Melanin is a pigment that is meant to scatter light for the purpose of protection against harmful radiation, and by a similar mechanism will attenuate light that arises from within the animal. Thus skin, fur and hair color may interfere with BL output and influence sensitivity of imaging. Studies have shown that the light emission from dark-colored mice such as the Rag2M strain, is significantly reduced when compared with white or hairless mice. For this reason, albino nude animals are often used in BL studies (Edinger, Cao et al. 2002). Curtis et al showed that even local depilation can cause pigment changes which interfere with BLI (Curtis, Calabro et al. 2010). Hemoglobin is another pigment that quenches light, thus highly vascularized organs tend to have lower levels of photon emission compared with less vascularized tissues. Finally, in the context of metastatic cancers or gene expression studies, the detection of smaller signals will depend on the presence of larger signals located close by as signal intensity from one region can attenuate less intense signals from other regions.

Three main aspects of cell physiology have been shown to affect BLI. First, Czupryna et al suggested that F-Luc activity can be substantially altered in studies where reactive oxygen species are elevated (Czupryna and Tsourkas 2011). This poses a very relevant problem in studies of tumor biology as oxidative stress can occur within a tumor or as a result of therapy. Second, hypoxic regions within tumors may also affect signal intensity. BL is dependent on oxygen and a number of studies have found that the amount of light emitted from luciferase-labeled cells is reduced as the oxygen concentration decreases (Cecic, Chan et al. 2007) (Moriyama, Niedre et al. 2008). Lastly, the expression level of ABC transporters can affect BLI intensity. Huang et al did a comparative study looking at the effects of different ATP-binding cassette (ABC) transporters on BLI readout when Click Beetle, Firefly, Renilla or Gaussia substrates were used in vitro. They show that ABCG2/BCRP is able to pump D-luciferin out of cells. Some groups have begun looking into increasing the stability of luciferin in vivo. For example, to improve the stability of and provide a continuous and prolonged delivery of the substrate D-luciferin for BLI, Kheirolomoom et al created a liposomal formulation of luciferin which had a prolonged release over 24 hours compared to the free form (Kheirolomoom, Kruse et al. 2010).

5. Use of BLI in oncology research

In oncology research, BL was first used in vitro and in vivo to assess cellular metabolism and indirectly to measure viability in an experiment known as the ATP assay. In this assay, luciferin and luciferase are added to media and in the presence of ATP, luciferin is oxidized giving off light that can be measured using a luminometer or a CCD camera. These assays are a direct measure of cell viability and can be used to assess cell proliferation and cytotoxicity in vitro (Garewal, Ahmann et al. 1986; Kuzmits, Aiginger et al. 1986; Kuzmits, Rumpold et al. 1986; Ahmann, Garewal et al. 1987; Sevin, Peng et al. 1988; Petru, Sevin et al. 1990; Crouch, Kozlowski et al. 1993). Using the ATP assay Garewal et al were able to distinguish between cytostatic and cytotoxic effects of therapeutics on a colon cancer cell line in vitro (Garewal, Ahmann et al. 1986).

In the late 80s and early 90's Mueller-Klieser and Walenta et al began mapping metabolites in excised tumor tissues using similar principles to the ATP assay. Their method allowed for the assessment of glucose, lactate, and ATP distributions in sections of tumors and normal tissue from cryobiopsies. Briefly, the procedure involves application of a solution containing gelatin and enzymes linked to luciferase. Upon tissue thawing, the enzyme solution diffuses into the tissue section initiating the BL reaction wherever the substrate of interest is found. The photons emitted are visualized in tumor sections using a CCD equipped microscope (Mueller-Klieser, Walenta et al. 1988; Mueller-Klieser, Kroeger et al. 1991; Walenta, Schroeder et al. 2002). Figure 4 shows a bright-field image (A) and BLI of ATP (B) from a cryosection of melanoma from the Syrian golden hamster. The BLI clearly indicates high concentrations of ATP in viable cell regions of the periphery. Furthermore, the studies show that high levels of glucose are found in the tumor periphery while necrotic tumor centers exhibited high lactate levels (Walenta, Dellian et al. 1992).

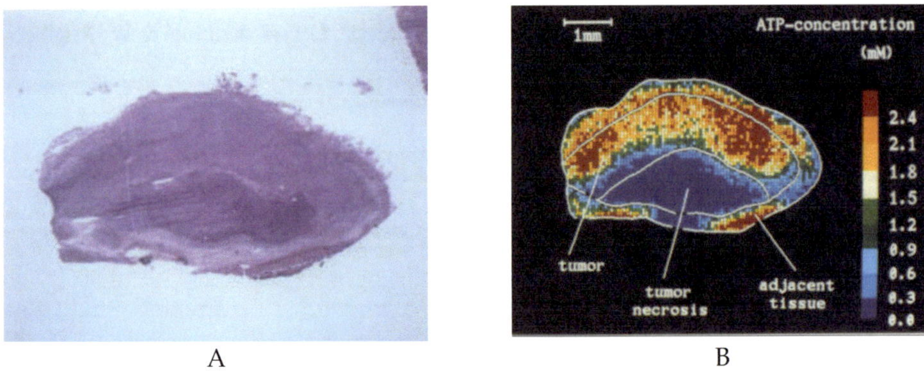

Fig. 4. Mapping ATP in melanoma of the Syrian Hamster using BLI. Cryosection stained with hematoxylin and eosin (A) shows necrosis in the center of the tumor, Colour-coded intensity image of bioluminescence (B) illustrates the local distrubiton of ATP concentrations. (Walenta, S., M. Dellian, et al. (1992). "Pixel-to-pixel correlation between images of absolute ATP concentrations and blood flow in tumors." Br J Cancer 66(6): 1099-1102. Reproduced by permission.)

The metabolic switch to aerobic glycolysis and enhanced lactate production is characteristic for aggressive tumor cells and a factor for tumor response and treatment outcome. Thus, BL mapping of metabolites can be used as an early marker for treatment response. Broggini-Tenzer et al use BL metabolite mapping strategies in mice carrying tumor xenografts derived from A549 lung cancer cells to show that metabolite levels are influenced by treatment with the microtubule stabilizing agent patupilone, ionizing radiation or a combination of the two modalities (Broggini-Tenzer, Vuong et al. 2011). Their results are shown in Figure 5. The BLI of tumor sections indicated that lactate levels were significantly reduced and glucose levels drastically increased in treated tumors compare to the untreated tumors. However, ATP levels did not change significantly with any of the treatments used.

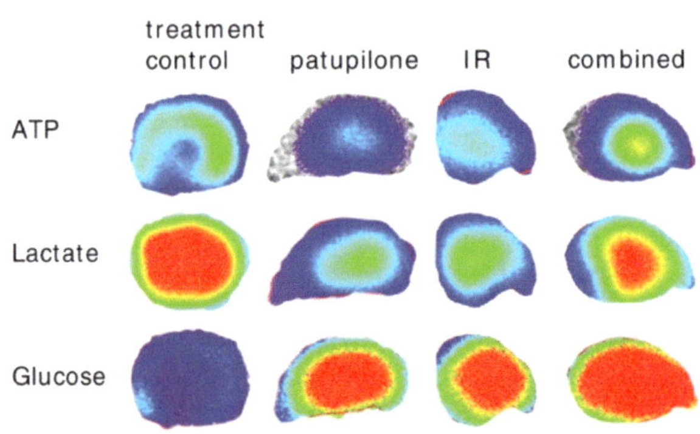

Fig. 5. Distribution of metabolites within lung tumor sections after treatment. Representative tumors are shown for each treatment group and consecutive sections were stained for ATP, lactate and glucose. Treatment with patupilone or irradiation exhibit decreased ATP levels and Lactate levels but increased glucose levels within the tumor core. (Broggini-Tenzer, A., V. Vuong, et al. (2011). "Metabolism of tumors under treatment: mapping of metabolites with quantitative bioluminescence." Radiother Oncol 99(3): 398-403.Reproduced by permission.)

The non-invasive utility of BLI as a small-animal imaging modality has led to the development of a wide range of luciferase positive orthotopic and metastatic animal models. Orthotopic models for breast cancer (Garcia, Jackson et al. 2008; Shan, Wang et al. 2008; Kalra, Warburton et al. 2009; Kalra, Anantha et al. 2011), bladder cancer (Mugabe, Matsui et al. 2011; van der Horst, van Asten et al. 2011), hepatocellular carcinoma (Frampas, Maurel et al. 2011), ovarian cancer (Cordero, Kwon et al. 2010; Bevis, McNally et al. 2011), head and neck SCC (Sano, Matsumoto et al. 2011), multiple myeloma (Runnels, Carlson et al. 2011), pancreatic cancer (Angst, Chen et al. 2010; McNally, Welch et al. 2010; Muniz, Barnes et al. 2011), glioma (Prasad, Sottero et al. 2011), lung cancer(Madero-Visbal, Colon et al. 2010; Li, Torossian et al. 2011; Yan, Xiao et al. 2011), prostate cancer (Svensson, Haverkamp et al. 2011), mesothelioma (Feng, Zhang et al. 2011), neuroblastoma (Teitz, Stanke et al. 2011; Tivnan, Tracey et al. 2011), rectal cancer (Huerta, Gao et al. 2011), renal cancer (Karam, Mason et al. 2003), sarcoma (Vikis, Jackson et al. 2010) have been established in vivo. Furthermore, several of these models have been used to develop systemic disease. For example, Mishra et al use intracardiac and intratibial inoculation of luciferase positive prostate cancer cells to investigate the effect of inhibiting TGFβ on osteoblastic tumor growth and incidence in vivo (Mishra, Tang et al. 2011). As seen below, our group used a luciferase positive human breast cancer cell line (LCC6-Luc) to establish orthotopic disease via mammary fatpad injections, ascitic disease via intraperitoneal injection and metastatic disease via intracardiac inoculation (Kalra, Anantha et al. 2011).

5.1 Capturing minimal disease as well as quantifying tumor development and growth

Luciferase expressing cells can be used in vivo to monitor tumor growth. In an animal model, it is possible to observe small numbers of luciferase positive cells following luciferin administration. Indeed, post-inoculation, luciferase activity can be detected allowing for the confirmation of cell injection. Because small numbers of cells are readily detected, quantitative measurements of disease burden can be done earlier and for the identification of metastatic spread. A study done in our laboratory used LCC6-Luc cells to inoculate animals orthotopically, intracardiac, or intraperitoneal, to establish mammary tumors, systemic disease and ascities disease respectively. The results are summarized in Figure 6. This study demonstrated that the growth of orthotopic, systemic, and ascities disease can be monitored from day zero upon tumor cell inoculation, through to day 28 where there is established disease. The BLI data was quantified and used to create growth curves for each model. Further, the information provided in Figure 1 was used to estimate the number of cells detected at day 7, 14, 21 and 28. These data, in turn, could be used to generate a tumor growth curve (Kalra, Anantha et al. 2011).

As suggested by the above example, a major advantage of using BLI is the sensitivity of capturing photon emissions from small numbers of cells and monitoring the onset of disease, tumor growth, primary tumor dynamics and progression to metastatsis. Using BLI it is possible to quantify the kinetics of tumor growth since photon emissions increase in proportion to the number of cells and thus disease burden. Light measurements can be made from whole body scans or from a selected region of interest and are most commonly quantified as total photon counts (photons/s). One of the first in vivo experiments performed using BLI of luciferase labeled cells was done by Edinger et al in 1999. This group used a Luciferase positive human cervical carcinoma cancer cell line (HeLa-Luc) to inoculate animals via subcutaneous, intraperitoneal and intravenous injection. They were able to visualize cells immediately following inoculation using all of the injection routes. According to this study, 1×10^3 cells could be detected in the peritoneal cavity, 1×10^4 at subcutaneous sites, and 1×10^6 circulating cells following injection (Edinger, Sweeney et al. 1999).

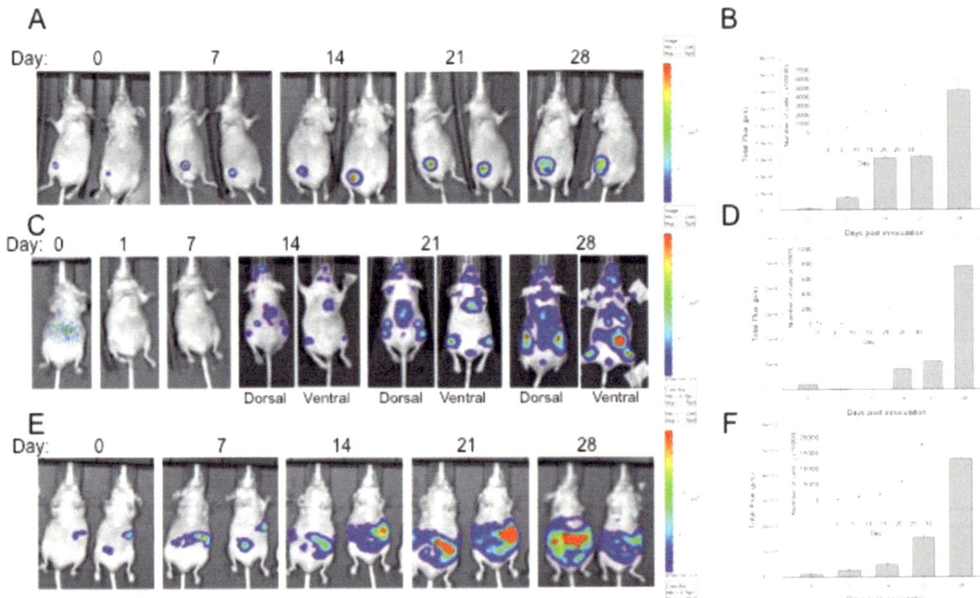

Fig. 6. Using Luciferase positive to establish orthotopic, systemic and ascitic tumors in animals. $LCC6^{WT\text{-}Luc}$ cells were inoculated orthotopically (A & B), via intracardiac injection (C & D), or intraperitoneally (E & F). BLI was used to monitor tumor growth. Images shown were acquired on days 0, 1, 7, 14, 21, and 28. Photon emissions were measured using whole body scans for each animal (representative images shown in A, C and E). BL data was quantified to generate growth curves (B, D and F). The inset graphs represent BLI data as it correlates to the number of cells. (Kalra, J., M. Anantha, et al. (2011). "Validating the use of a luciferase labeled breast cancer cell line, MDA435LCC6, as a means to monitor tumor progression and to assess the therapeutic activity of an established anticancer drug, docetaxel (Dt) alone or in combination with the ILK inhibitor, QLT0267." Cancer Biol Ther 11(9): 826-838. Reproduced by permission from the author.)

In 2003 Jenkins et al developed a luciferase positive prostate cancer, lung cancer and colon cancer model to study tumor growth in vivo. Bioluminescent PC-3M-Luc-C6 human prostate cancer cells were implanted subcutaneously into mice and were monitored for tumor growth using BLI. They show that BLI data correlated to standard external caliper measurements of tumor volume, but BLI permitted earlier detection of tumor cells. In the lung colonization cancer model, bioluminescent A549-Luc-C8 human lung cancer cells were injected intravenously and lung metastases were successfully monitored in vivo by whole animal imaging. Bioluminescent HT-29-Luc-D6 human colon cancer cells implanted subcutaneously produced metastases to lung and lymph nodes. Both primary tumors and micrometastases were detected by BLI in vivo (Jenkins, Oei et al. 2003).

More recently in 2011, van der Horst et al used a re-engineered firefly luciferase (Luciferase 2) in which a transcription-factor binding sites that compromised luciferase expression was removed. A Luciferase 2-positive human transitional carcinoma cell line (UM-UC-3-Luc) was used to inoculate either the bladder of mice to produce an orthotopic model with systemic metastases or in the left cardiac ventricle to develop a model that simulates bone metastasis. This group was able to detect 100 cells three hours after subcutaneous inoculation. They were also able to detect micrometastases from both the orthotopic and metastatic tumors (van der Horst, van Asten et al. 2011).

In 2002, Bhaumik et al show that D-luciferin (the substrate for F-Luc) does not serve as a substrate for Renilla Luciferase (R-Luc), and coelenterazine (the substrate for R-Luc) does not serve as a substrate for F-Luc in cell culture or in living mice. This group made stable transfections of a rat glioma cell line (C6) with R-Luc or F-Luc. Mice were inoculated with C6-R-Luc in the left forearm, and C6-F-Luc in the right forearm. Once tumors had established, animals were subject to BLI using D-luciferin or coelenterazine. As shown in Figure 7, D-luciferin delivery was associated with photons emitted from the tumor cells in the right forearm, while coeleterazine was associated with photons emitted from the left forearm. Elegantly, Bhaumik et al were able to demonstrate that both R-Luc and F-Luc expression can be imaged in the same living mouse. This pivotal finding adds an extra layer of complexity to the BLI modality in that different luciferins can be used to track at least two separate 1) molecular events, 2) cell populations such as stem cells versus tumor cells, 3) gene therapy vectors, or 4) endogenous genes through the use of two reporter luciferase genes (Bhaumik and Gambhir 2002).

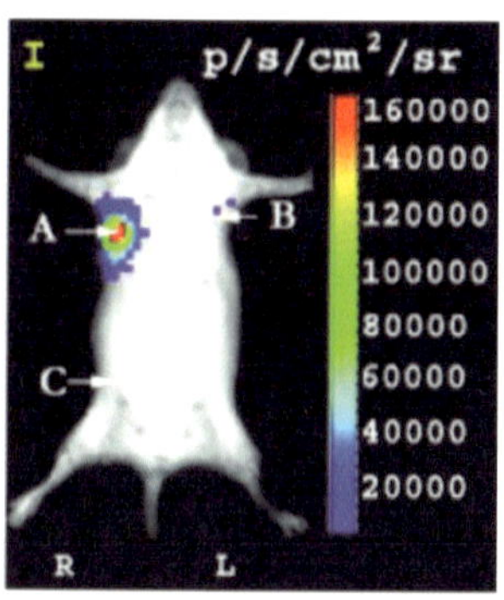

A

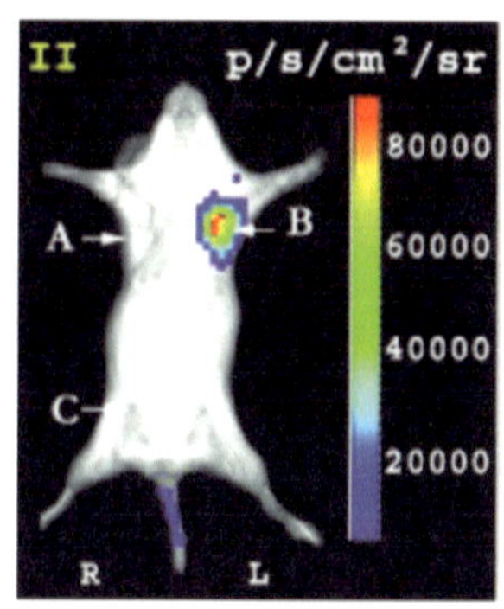

B

Fig. 7. BLI of F-Luc and R-Luc activity in the same animal. Both C6-F-Luc (A) and C6-R-Luc (B) cells were implanted subcutaneously at the right or left forearm sites respectively in the same mouse. Injection of D-luciferin via tail-vein in the mouse in figure 4A shows bioluminescence from site A and minimal signal from site B. Injection of coelenterazine via tail-vein in the mouse in Figure 3B exhibit bioluminescence from site B but minimal signal from site A. (Bhaumik, S. and S. S. Gambhir (2002). "Optical imaging of Renilla luciferase reporter gene expression in living mice." 99 (1): 377-382. Copyright (2002) National Academy of Sciences, U.S.A. Reproduced by permission.)

Indeed using Bhaumik's findings, Wang et al separately labeled murine breast cancer cells (4T1) with an R-Luc-monomeric red fluorescence protein (R-Luc-mRFP) reporter vector and mesenchymal stem cells (MSC) with a F-Luc-enhanced green fluorescence protein (F-Luc-eGFP) reporter vector in order to study how MSC traffic and differentiate in either subcutaneous or metastatic animal models. Wang et al were successfully able to monitor tumor growth by R-Luc BLI and the MSC's by F-Luc BLI in the same animal (Wang, Cao et al. 2009).

As already indicated, F-Luc requires ATP in order to produce light, thus only metabolically active and oxygen rich cells contribute to the signal observed in BLI. A decrease in signal intensity occurs as cells undergo apoptosis or necrosis. In one of our own studies, we used the highly aggressive breast cancer cell line MDA MB 435/LCC6 to make a BL orthotopic mouse model. This cell line is known to rapidly develop tumors with necrotic cores. Using BLI it was noted that 28 days post tumor inoculation, the center of the tumor no longer emitted BL photons suggestive of a metabolically inactive tumor core. Dead or necrotic regions within a tumor, would still contribute to its volume, therefore traditional caliper measurements would have provided inaccurate results.

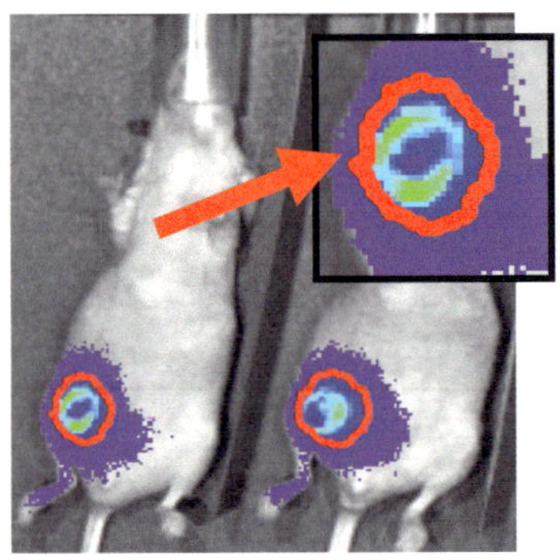

Fig. 8. Orthotopic LCC6-Luc tumors have necrotic cores that can be visualized using BLI. LCC6-Luc cells were inoculated into the mammary fat pad of female mice. Tumors were monitored using BLI. On day 28 post tumor cell inoculation, non-luminescing regions within the tumor was visualized after BLI suggestive of necrotic centers. (Kalra, J., C. Warburton, et al. (2009). "QLT0267, a small molecule inhibitor targeting integrin-linked kinase (ILK), and docetaxel can combine to produce synergistic interactions linked to enhanced cytotoxicity, reductions in P-AKT levels, altered F-actin architecture and improved treatment outcomes in an orthotopic breast cancer model." Breast Cancer Res 11(3): R25. Reproduced by permission.)

Since only metabolically active cells may produce BL, BLI can be used to indicate a positive drug response. BLI can be used to measure cell death with treatment using cytotoxic agents or even reduced proliferation post treatment with cytostatic agents. Many studies have been published to date describing the use of BLI to determine the efficacy of a variety of drugs currently being used in clinic and to look for valuable new drugs in the development pipeline. Because BLI is able to capture minimal disease, treatment success and cure of animals can be determined at earlier stages of, at metastasis and to monitor relapse before any clinical signs of disease are detectable. Further, since animals are monitored non-invasively, serial sacrifice of groups of animals at intermediate time-points is not necessary. This saves on the number of animals used per study and also means that multiple parameters of drug efficacy and dosing can be studied simultaneously. Li et al used luciferase-expressing A549 cancer cells injected into the mediastinum of athymic nude mice to determine whether the luciferase positive model would allow for monitoring of response to therapeutic interventions. Animals were treated with paclitaxel or irradiated, and tumor burden was monitored using BLI. They noted that tumors responded to paclitaxel or radiation as shown by decreased tumor BL which also correlated to improved overall survival (Li, Torossian et al. 2011). Mugabe et al, use BLI to determine if a mucoadhesive nanoparticulate form of docetaxel is able to improve treatment of a bladder cancer by increasing the dwell time and uptake of the intravesical drug (Mugabe, Matsui et al. 2011). In 2009, Graeser et al use BLI to show that a liposomal formulation of gemcitabine has improved anti-tumor and anti-metastatic effects in an orthotopic model of pancreatic cancer when compared to the free drug (Graeser, Bornmann et al. 2009). As most agents are used as part of combination regimens, BLI is an ideal technique to study combination effects against single agent therapies and to elucidate combinatorial ratios and scheduling, as these experiments require multiple study arms and large numbers of animals. Prasad et al show that in a luciferase positive glioma model, combining a cytostatic agent (PI3K/mTOR dual inhibitor (XL765)) with a clinically relevant agent, temozolomide (TMZ), resulted in an additive reduction in tumor BL compared with control. The BLI data correlated to improvement in median survival time in the combination treated group (Prasad, Sottero et al. 2011). Finally, using a novel approach to study pharmacodynamics, Pensel et al use two imaging modalities to study probe accumulation at the site of tumor tissue. This group used a luciferase positive human leukemia model (HL-60-Luc) and a radiolabeled probe (spermine), imaged with BLI and SPECT respectively, to demonstrate that the spermine conjugate accumulates in tumor cells (Pesnel, Guminski et al. 2011).

5.2 Investigating key cancer processes in vivo

In addition to monitoring tumor initiation and growth as indicated above, BLI has been used in a variety of mechanistic studies. For example, in 2002 Shuetz et al were able to track transcription of luciferase reporter genes in an in vivo model of liver cancer before and after treatment (Schuetz, Lan et al. 2002). In 2003 Luker et al used a ubiquitin luciferase reporter to follow protesomal function in vivo before and after treatment with proteosome inhibitors (Luker, Pica et al. 2003). In oncology research it is now possible to devise reporter strategies to assess key cancer processes such as dysregulated signaling, induction of apoptosis and angiogeneis in vivo.

5.2.1 Imaging apoptosis

Imaging apoptosis in vivo using a non-invasive modality would be a valuable method to evaluate drugs that induce programmed cell death. To this end, Laxman et al constructed an apoptosis biosensor by fusing the estrogen regulatory (ER) domain to F-Luc. The ER domain is able to silence the enzymatic activity of luciferase. The construct was further engineered so that the luciferase protein was flanked by the protease cleavage site for caspase-3. This cleavage site consists of aspartic acid (D), glutamic acid (E), valine (V), and aspartic acid (D) and is known as DEVD. If the DEVD site is cleaved by caspase-3, luciferase would be released from the construct, and the silencing effect from ER would be ablated. Stable human glioma cells line expressing this luciferase construct were generated and inoculated into animals subcutaneously. Animals were treated with tumor necrosis factor α-related apoptosis-inducing ligand (TRAIL), which induces apoptosis. With activation of caspase-3 the DEVD sites were cleaved, luciferase was able to fold appropriately and upon exposure to luciferin, BL photons were produced. Therefore, apoptosis was successfully imaged non-invasively using BLI (Laxman, Hall et al. 2002). Using another methodology, Niers et al engineered the naturally secreted G-Luc so that it is separated by the DEVD sequence. They showed that this fusion protein was retained in the cytoplasm of transfected cells in an inactive form. Upon induction of apoptosis, the DEVD peptide was cleaved in response to caspase-3 activation, freeing G-Luc, which then entered the secretory pathway where it was folded properly and released from the cells. The G-Luc can be detected in the conditioned medium in culture or in blood from live animals (Niers, Kerami et al. 2011). Scabini et al 2011 use a similar approach however in this case a formulated Z-DEVD-aminoluciferin is delivered intraperiotneal to mice carrying human colon cancer or human glioblastoma cell lines engineered to express luciferase. Upon induction of apoptosis Z-DEVD-aminoluciferin is cleaved by caspase 3/7 releasing aminoluciferin that is now free to react with luciferase to generate measurable BL. This group was able to show that after camptothecin and temozolomide treatment of xenograft mouse models of colon cancer and glioblastoma respectively, the treated mice showed higher induction of Z-DEVD-aminoluciferin luminescent signal when compared to the untreated group. Combining D-luciferin that measures the total tumor burden, with Z-DEVD-aminoluciferin that assesses apoptosis induction via caspase activation, they were able to relate inhibition of tumor growth with induction of apoptosis after treatment in the same animal over time (Scabini, Stellari et al. 2011). Hickson et al use the same methodology in a luciferase positive ovarian cancer and breast cancer model. In these experiments, tumor cells were inoculated and allowed to establish, subsequently animals were treated with docetaxel. Animals were injected with the Z-DEVD-aminoluciferin before BL images were acquired. This group shows that more light was detected in the docetaxel-treated group compared with the untreated group (Hickson, Ackler et al. 2010).

5.2.2 Imaging tumor hypoxia and angiogenesis

Oxygen is needed for proper cellular metabolism, thus hypoxia, which is common in proliferating cancers, can significantly alter tumor biology on a molecular level. Monitoring hypoxia in vivo can provide important information on tumor biology and response to treatment. The transcription factor Hypoxia-inducing factor 1 (HIF1), is induced under conditions of hypoxia and specifically binds to the hypoxia response element (HRE) to promote transcriptional activation. Reporter vectors based on HRE elements driving luciferase expression have been designed for longitudinal imaging of hypoxia. For example, Viola et al inoculated mice with breast carcinoma cells transfected with an HIF-1α luciferase reporter construct and treated these animals using cyclophosphamide or paclitaxel. They showed that cyclophosphamide significantly inhibited tumor growth and caused an increase in HIF-1α protein levels as quantified using BLI (Viola, Provenzale et al. 2008). As discussed above, a transgenic mouse model was generated in which a chimeric protein consisting of HIF-1α oxygen-dependent degradation domain (ODD) is fused to luciferase. Hypoxic stress lead to the accumulation of ODD-luciferase which could then be identified by non-invasive BL measurement (Goldman, Chen et al. 2011).

Hypoxia stimulates secretion of vascular endothelial growth factor (VEGF) which in turn promotes angiogenesis. Transgenic mice have been engineered to express the VEGF receptor

2 (VEGFR2) promoter that drives F-Luc expression. This mouse model can be used to monitor angiogenesis induced by tumors. Angst et al sought to investigate pancreatic cancer angiogenesis and thus employed the VEGFR2-Luc mouse. After orthotopic inoculation of pancreatic cells, light emission corresponding to VEGFR activity began at day 4, which this group suggests is likely due to wound healing, and continued throughout the experimental period during tumor growth suggesting angiogenesis was occurring. The BL results were confirmed using immunohistochemical staining for CD31 (Angst, Chen et al. 2010). In 2007, Faley et al generated a transgenic reporter mouse, VEGF-GFP/Luc, in which an enhanced green fluorescent protein-luciferase fusion protein is expressed under the control of a human VEGF-A promoter. The VEGF-GFP/Luc animals exhibited intense BL throughout the body at 1 week of age, but the signals declined as the mice grew so that the adult VEGF-GFP/Luc mouse showed BL only in areas undergoing active wound healing. However, in VEGF-GFP/Luc/MMTV mice, BL is observed in spontaneous tumors indicative of active angiogenesis (Faley, Takahashi et al. 2007).

5.2.3 Imaging Protein – Protein interactions and cell signalling

In order to have a mechanistic understanding of tumor biology and response to therapy, oncology research focuses on molecular alterations in the tumor or microenvironment. Under many circumstances up-regulation of oncogenes results in changes in protein–protein interactions, alterations in kinase activity and associated changes in important signalling pathways that promote tumour cell survival and proliferation. Much work has been accomplished to study these signalling cascades in vitro and in ex vivo tissue samples and as a result many therapies have been developed to target these dysregulated pathways. For these reasons there has been a great deal of interest in developing methods to visualize molecular changes in live animals.

Three general methods are currently available for imaging protein-protein interactions in living subjects using reporter genes: a modified mammalian two-hybrid system, a bioluminescence resonance energy transfer (BRET) system, and split reporter protein complementation and reconstitution strategies, these methods were reviewed by Massoud et al in 2007 (Massoud, Paulmurugan et al. 2007). Paulmurgan developed the split reporter system in vivo using very strongly interacting proteins MyoD and Id (Paulmurugan, Umezawa et al. 2002). In 2004 this same group used split synthetic R-Luc protein to evaluate heterodimerization of FRB and FKBP12 mediated by rapamycin. The rapamycin-mediated dimerization of FRB and FKBP12 was studied in living mice by locating, quantifying, and timing the R-Luc BL. Their work demonstrates that the split reporter system can be used to screen small molecule drugs that impact protein-protein interactions in living animals (Paulmurugan, Massoud et al. 2004).

It is also possible to use BLI for the evaluation of enzymatic activity such as kinase activity, in vivo. Khan et al established a luciferase-based reporter to image EGFR kinase activity in an in vivo model of squamous cell carcinoma (SCC). The EGFR Kinase reporter (EKR) is a multidomain chimeric reporter where BL can be used as a marker for EGFR kinase activity. The reporter is phosphorylated in the presence of active EGFR which interferes with luciferase activity, if the substrate is not phosphorylated BL is available for imaging. This reporter can therefore be used as an indicator for EGFR inhibition. Khan et al demonstrated that a small molecule inhibitor of EGFR kinase activity (erlotinib) was able to inhibit kinase activity in the SSC tumor model using BLI (Khan, Contessa et al. 2011).

BLI has also been used to monitor cell cycle signaling. In vivo BLI can be used to visualize the accumulation of p27-Luc in human tumor cells after the administration of Cdk2 inhibitory drugs (Zhang and Kaelin 2005). Briat et al have generated luciferase-based p53-reporter animals to monitor p53 activation. They showed that in response to doxorubicin induced DNA damage, female animals had weak p53 luciferase activity in the oral cavity while in males, the signal increased in the lower abdominal region (Briat and Vassaux 2008). A reporter molecule has also been developed to measure Akt activity in animals via BLI. The reporter comprises of an engineered luciferase molecule that undergoes a conformational change and gains functionality in response to phosphorylation by Akt (Zhang, Lee et al. 2007).

6. BLI in the study of gene activity, delivery and silencing

BLI provides a means to study gene delivery, activation using inducible systems, or silencing of tumor promoting genes using RNA interference (RNAi). Delivery of genes can be accomplished using multiple strategies, such as bacterial or viral vector delivery systems, immune cell and stem cell based delivery systems or encapsulation using special nanoparticle formulations such as liposomes or glucosylated polyethyleneimine. Monitoring gene delivery using BLI has also been accomplished. For example Hu et al were able to monitor TGF β receptor gene therapy efficacy in luciferasc positive breast cancer metastases simply by monitoring metastases development after gene delivery (Hu, Gerseny et al. 2011). BLI also enables the evaluation of delivery itself. For example, Badr et al have made a construct that comprises of 1) G-Luc, 2) the therapeutic gene cytosine deaminase and 3) uracil phosphoribosyltransferase which converts the nontoxic compound 5-fluorocytosine (5FC) into the drug 5-fluorouracil. A glioma cell line was engineered to express F-Luc. When the constructed gene transfers into tumors, G-Luc allows monitoring of the duration and magnitude of transgene expression while F-Luc imaging was used to monitor tumor growth and response to therapy with the pro-drug 5FC (Badr, Niers et al. 2011). Ahn et al made an adenoviral vector construct where the Survivin promoter (pSurv) amplifies the expression of both the reporter gene F-Luc and therapeutic gene TRAIL. In an orthotopic hepatocellular carcinoma (HCC) rat model, they showed that after systemic administration of the vector, BLI revealed increased F-Luc activity within the tumor compared with the liver indicating that the vector shows tumor-specific transgene expression (Ahn, Ronald et al. 2011). From a gene silencing standpoint, use of luciferase-targeting siRNAs has been studied to define the proof of principle that lipid based systemic administration of luciferase targeting siRNA is able to silence luciferase gene expression in glioma (Ofek, Fischer et al. 2010) and bone metastases (Takeshita, Hokaiwado et al. 2009).

7. Conclusion

BLI is a well-established tool in cancer research that can provide valuable insight into biological processes in intact cells, excised tissues as well as in animal models of cancer. It can facilitate medium-throughput assessments, it is very sensitive, and reasonably non-invasive. The utility of BLI surpasses simple surveying of tumor growth. More specifically, BLI can be used in the development of sophisticated animal models that examine minimal or metastatic disease, therapeutic efficacy, disease relapse, mechanistic assessments of new treatment regimens, protein-protein interactions, and to gain a better understanding of basic cancer biology. BLI facilitates visualization of processes such as metastasis, angiogenesis, apoptosis and cell signaling in vivo. As noted by Badr et al, the sensitivity of BLI allows for the early detection of tumors and therefore can be useful in the design of preclinical studies assessing prevention strategies (Badr and Tannous 2011). As the BLI modality becomes more popular, work is being done to improve the technology in order to optimize the sensitivity and detection of BL photons. For example, IVIS by Caliper has introduced a system where CT scans and BLI can be used simultaneously to generate three-dimensional images of animals and their disease. Other groups are working on engineering novel luciferases and luciferins to enhance their stability and pharmacokinetics in vivo. As indicated, it is recognized that BLI faces some challenges (distribution and absorption of the substrate as well as scattering issues effecting quantification), however continued use of BLI and proper preclinical study design can overcome most of the problems associated with this modality. BLI as a small animal imaging modality will be an integral part of the future of pre-clinical oncology research and its applications are being refined to achieve an understanding of disease development and response to therapy that was not previously possible.

8. References

Ahmann, F. R., H. S. Garewal, et al. (1987). "Intracellular adenosine triphosphate as a measure of human tumor cell viability and drug modulated growth." *In Vitro Cell Dev Biol* 23(7): 474-480.

Ahn, B. C., J. A. Ronald, et al. (2011). "Potent, tumor-specific gene expression in an orthotopic hepatoma rat model using a Survivin-targeted, amplifiable adenoviral vector." *Gene Ther* 18(6): 606-612.

Angst, E., M. Chen, et al. (2010). "Bioluminescence imaging of angiogenesis in a murine orthotopic pancreatic cancer model." *Mol Imaging Biol* 12(6): 570-575.

Badr, C. E., J. M. Niers, et al. (2011). "Suicidal gene therapy in an NF-kappaB-controlled tumor environment as monitored by a secreted blood reporter." *Gene Ther* 18(5): 445-451.

Badr, C. E. and B. A. Tannous (2011). "Bioluminescence imaging: progress and applications." *Trends Biotechnol.*

Baert, A. L. (2008). *Encyclopedia of Diagnostic Imaging*, Springer Reference.

Bevis, K. S., L. R. McNally, et al. (2011). "Anti-tumor activity of an anti-DR5 monoclonal antibody, TRA-8, in combination with taxane/platinum-based chemotherapy in an ovarian cancer model." *Gynecol Oncol* 121(1): 193-199.

Bhaumik, S. and S. S. Gambhir (2002). "Optical imaging of Renilla luciferase reporter gene expression in living mice." *Proc Natl Acad Sci U S A* 99(1): 377-382.

Biron-Pain, K. and Y. St-Pierre (2011). "Monitoring mmp-9 gene expression in stromal cells using a novel transgenic mouse model." *Cell Mol Life Sci.*

Briat, A. and G. Vassaux (2008). "A new transgenic mouse line to image chemically induced p53 activation in vivo." *Cancer Sci* 99(4): 683-688.

Broggini-Tenzer, A., V. Vuong, et al. (2011). "Metabolism of tumors under treatment: mapping of metabolites with quantitative bioluminescence." *Radiother Oncol* 99(3): 398-403.

Cecic, I., D. A. Chan, et al. (2007). "Oxygen sensitivity of reporter genes: implications for preclinical imaging of tumor hypoxia." *Mol Imaging* 6(4): 219-228.

Cordero, A. B., Y. Kwon, et al. (2010). "In vivo imaging and therapeutic treatments in an orthotopic mouse model of ovarian cancer." *J Vis Exp*(42).

Crouch, S. P., R. Kozlowski, et al. (1993). "The use of ATP bioluminescence as a measure of cell proliferation and cytotoxicity." *J Immunol Methods* 160(1): 81-88.

Curtis, A., K. Calabro, et al. (2010). "Temporal Variations of Skin Pigmentation in C57Bl/6 Mice Affect Optical Bioluminescence Quantitation." *Mol Imaging Biol.*

Czupryna, J. and A. Tsourkas (2011). "Firefly luciferase and RLuc8 exhibit differential sensitivity to oxidative stress in apoptotic cells." *PLoS One* 6(5): e20073.

de Wet, J. R., K. V. Wood, et al. (1987). "Firefly luciferase gene: structure and expression in mammalian cells." *Mol Cell Biol* 7(2): 725-737.

Dickson, P. V., B. Hamner, et al. (2007). "In vivo bioluminescence imaging for early detection and monitoring of disease progression in a murine model of neuroblastoma." *J Pediatr Surg* 42(7): 1172-1179.

Dussmann, P., J. I. Pagel, et al. (2011). "Live in vivo imaging of Egr-1 promoter activity during neonatal development, liver regeneration and wound healing." *BMC Dev Biol* 11: 28.

Edinger, M., Y. A. Cao, et al. (2002). "Advancing animal models of neoplasia through in vivo bioluminescence imaging." *Eur J Cancer* 38(16): 2128-2136.

Edinger, M., T. J. Sweeney, et al. (1999). "Noninvasive assessment of tumor cell proliferation in animal models." *Neoplasia* 1(4): 303-310.

El-Deiry, W. S., C. C. Sigman, et al. (2006). "Imaging and oncologic drug development." *J Clin Oncol* 24(20): 3261-3273.

Faley, S. L., K. Takahashi, et al. (2007). "Bioluminescence imaging of vascular endothelial growth factor promoter activity in murine mammary tumorigenesis." *Mol Imaging* 6(5): 331-339.

Feng, M., J. Zhang, et al. (2011). "In vivo imaging of human malignant mesothelioma grown orthotopically in the peritoneal cavity of nude mice." *J Cancer* 2: 123-131.

Frampas, E., C. Maurel, et al. (2011). "The intraportal injection model for liver metastasis: advantages of associated bioluminescence to assess tumor growth and influences on tumor uptake of radiolabeled anti-carcinoembryonic antigen antibody." *Nucl Med Commun* 32(2): 147-154.

Garcia, T., A. Jackson, et al. (2008). "A convenient clinically relevant model of human breast cancer bone metastasis." *Clin Exp Metastasis* 25(1): 33-42.

Garewal, H. S., F. R. Ahmann, et al. (1986). "ATP assay: ability to distinguish cytostatic from cytocidal anticancer drug effects." *J Natl Cancer Inst* 77(5): 1039-1045.

Geusz, M. E., K. T. Blakely, et al. (2010). "Elevated mPer1 gene expression in tumor stroma imaged through bioluminescence." *Int J Cancer* 126(3): 620-630.

Goldman, S. J., E. Chen, et al. (2011). "Use of the ODD-luciferase transgene for the non-invasive imaging of spontaneous tumors in mice." *PLoS One* 6(3): e18269.

Graeser, R., C. Bornmann, et al. (2009). "Antimetastatic effects of liposomal gemcitabine and empty liposomes in an orthotopic mouse model of pancreatic cancer." *Pancreas* 38(3): 330-337.

Hickson, J., S. Ackler, et al. (2010). "Noninvasive molecular imaging of apoptosis in vivo using a modified firefly luciferase substrate, Z-DEVD-aminoluciferin." *Cell Death Differ* 17(6): 1003-1010.

Hsieh, C. L., Z. Xie, et al. (2005). "A luciferase transgenic mouse model: visualization of prostate development and its androgen responsiveness in live animals." *J Mol Endocrinol* 35(2): 293-304.

Hu, Z., H. Gerseny, et al. (2011). "Oncolytic Adenovirus Expressing Soluble TGFbeta Receptor II-Fc-mediated Inhibition of Established Bone Metastases: A Safe and Effective Systemic Therapeutic Approach for Breast Cancer." *Mol Ther* 19(9): 1609-1618.

Huerta, S., X. Gao, et al. (2011). "Murine orthotopic model for the assessment of chemoradiotherapeutic interventions in rectal cancer." *Anticancer Drugs* 22(4): 371-376.

Iyer, M., F. B. Salazar, et al. (2004). "Noninvasive imaging of enhanced prostate-specific gene expression using a two-step transcriptional amplification-based lentivirus vector." *Mol Ther* 10(3): 545-552.

Iyer, M., F. B. Salazar, et al. (2005). "Non-invasive imaging of a transgenic mouse model using a prostate-specific two-step transcriptional amplification strategy." *Transgenic Res* 14(1): 47-55.

Jenkins, D. E., Y. Oei, et al. (2003). "Bioluminescent imaging (BLI) to improve and refine traditional murine models of tumor growth and metastasis." *Clin Exp Metastasis* 20(8): 733-744.

Jia, W., S. Wang, et al. (2011). "A BAC transgenic reporter recapitulates in vivo regulation of human telomerase reverse transcriptase in development and tumorigenesis." *FASEB J* 25(3): 979-989.

Kalra, J., M. Anantha, et al. (2011). "Validating the use of a luciferase labeled breast cancer cell line, MDA435LCC6, as a means to monitor tumor progression and to assess the therapeutic activity of an established anticancer drug, docetaxel (Dt) alone or in combination with the ILK inhibitor, QLT0267." *Cancer Biol Ther* 11(9): 826-838.

Kalra, J., C. Warburton, et al. (2009). "QLT0267, a small molecule inhibitor targeting integrin-linked kinase (ILK), and docetaxel can combine to produce synergistic interactions linked to enhanced cytotoxicity, reductions in P-AKT levels, altered F-actin architecture and improved treatment outcomes in an orthotopic breast cancer model." *Breast Cancer Res* 11(3): R25.

Karam, J. A., R. P. Mason, et al. (2003). "Molecular imaging in prostate cancer." *J Cell Biochem* 90(3): 473-483.

Khan, A. P., J. N. Contessa, et al. (2011). "Molecular imaging of epidermal growth factor receptor kinase activity." *Anal Biochem* 417(1): 57-64.

Kheirolomoom, A., D. E. Kruse, et al. (2010). "Enhanced in vivo bioluminescence imaging using liposomal luciferin delivery system." *J Control Release* 141(2): 128-136.

Kuzmits, R., P. Aiginger, et al. (1986). "Assessment of the sensitivity of leukaemic cells to cytotoxic drugs by bioluminescence measurement of ATP in cultured cells." *Clin Sci (Lond)* 71(1): 81-88.

Kuzmits, R., H. Rumpold, et al. (1986). "The use of bioluminescence to evaluate the influence of chemotherapeutic drugs on ATP-levels of malignant cell lines." *J Clin Chem Clin Biochem* 24(5): 293-298.

Laxman, B., D. E. Hall, et al. (2002). "Noninvasive real-time imaging of apoptosis." *Proc Natl Acad Sci U S A* 99(26): 16551-16555.

Lee, Y. C., C. F. Huang, et al. (2010). "Src family kinase/abl inhibitor dasatinib suppresses proliferation and enhances differentiation of osteoblasts." *Oncogene* 29(22): 3196-3207.

Li, B., A. Torossian, et al. (2011). "A novel bioluminescence orthotopic mouse model for advanced lung cancer." *Radiat Res* 176(4): 486-493.

Li, F., Q. Cheng, et al. (2010). "Generation of a novel transgenic mouse model for bioluminescent monitoring of survivin gene activity in vivo at various pathophysiological processes: survivin expression overlaps with stem cell markers." *Am J Pathol* 176(4): 1629-1638.

Lin, A. H., J. Luo, et al. (2005). "Global analysis of Smad2/3-dependent TGF-beta signaling in living mice reveals prominent tissue-specific responses to injury." *J Immunol* 175(1): 547-554.

Lipshutz, G. S., D. Titre, et al. (2003). "Comparison of gene expression after intraperitoneal delivery of AAV2 or AAV5 in utero." *Mol Ther* 8(1): 90-98.

Luker, G. D., C. M. Pica, et al. (2003). "Imaging 26S proteasome activity and inhibition in living mice." *Nat Med* 9(7): 969-973.

Luo, J. and T. Wyss-Coray (2009). "Bioluminescence analysis of Smad-dependent TGF-beta signaling in live mice." *Methods Mol Biol* 574: 193-202.

Lyons, S. K., E. Lim, et al. (2006). "Noninvasive bioluminescence imaging of normal and spontaneously transformed prostate tissue in mice." *Cancer Res* 66(9): 4701-4707.

Madero-Visbal, R. A., J. F. Colon, et al. (2010). "Bioluminescence imaging correlates with tumor progression in an orthotopic mouse model of lung cancer." *Surg Oncol.*

Massoud, T. F., R. Paulmurugan, et al. (2007). "Reporter gene imaging of protein-protein interactions in living subjects." *Curr Opin Biotechnol* 18(1): 31-37.

McNally, L. R., D. R. Welch, et al. (2010). "KISS1 over-expression suppresses metastasis of pancreatic adenocarcinoma in a xenograft mouse model." *Clin Exp Metastasis* 27(8): 591-600.

Mishra, S., Y. Tang, et al. (2011). "Blockade of transforming growth factor-beta (TGFbeta) signaling inhibits osteoblastic tumorigenesis by a novel human prostate cancer cell line." *Prostate* 71(13): 1441-1454.

Momota, H. and E. C. Holland (2005). "Bioluminescence technology for imaging cell proliferation." *Curr Opin Biotechnol* 16(6): 681-686.

Moriyama, E. H., M. J. Niedre, et al. (2008). "The influence of hypoxia on bioluminescence in luciferase-transfected gliosarcoma tumor cells in vitro." *Photochem Photobiol Sci* 7(6): 675-680.

Moriyama, Y., E. H. Moriyama, et al. (2005). "In vivo study of the inflammatory modulating effects of low-level laser therapy on iNOS expression using bioluminescence imaging." *Photochem Photobiol* 81(6): 1351-1355.

Mueller-Klieser, W., M. Kroeger, et al. (1991). "Comparative imaging of structure and metabolites in tumours." *Int J Radiat Biol* 60(1-2): 147-159.

Mueller-Klieser, W., S. Walenta, et al. (1988). "Metabolic imaging in microregions of tumors and normal tissues with bioluminescence and photon counting." *J Natl Cancer Inst* 80(11): 842-848.

Mugabe, C., Y. Matsui, et al. (2011). "In vivo evaluation of mucoadhesive nanoparticulate docetaxel for intravesical treatment of non-muscle-invasive bladder cancer." *Clin Cancer Res* 17(9): 2788-2798.

Muniz, V. P., J. M. Barnes, et al. (2011). "The ARF tumor suppressor inhibits tumor cell colonization independent of p53 in a novel mouse model of pancreatic ductal adenocarcinoma metastasis." *Mol Cancer Res* 9(7): 867-877.

Niers, J. M., M. Kerami, et al. (2011). "Multimodal in vivo imaging and blood monitoring of intrinsic and extrinsic apoptosis." *Mol Ther* 19(6): 1090-1096.

Nyati, M. K., Z. Symon, et al. (2002). "The potential of 5-fluorocytosine/cytosine deaminase enzyme prodrug gene therapy in an intrahepatic colon cancer model." *Gene Ther* 9(13): 844-849.

O'Neill, K., S. K. Lyons, et al. (2010). "Bioluminescent imaging: a critical tool in pre-clinical oncology research." *J Pathol* 220(3): 317-327.

Ofek, P., W. Fischer, et al. (2010). "In vivo delivery of small interfering RNA to tumors and their vasculature by novel dendritic nanocarriers." *FASEB J* 24(9): 3122-3134.

Paulmurugan, R., T. F. Massoud, et al. (2004). "Molecular imaging of drug-modulated protein-protein interactions in living subjects." *Cancer Res* 64(6): 2113-2119.

Paulmurugan, R., Y. Umezawa, et al. (2002). "Noninvasive imaging of protein-protein interactions in living subjects by using reporter protein complementation and reconstitution strategies." *Proc Natl Acad Sci U S A* 99(24): 15608-15613.

Pesnel, S., Y. Guminski, et al. (2011). "(99m)Tc-HYNIC-spermine for imaging polyamine transport system-positive tumours: preclinical evaluation." *Eur J Nucl Med Mol Imaging* 38(10): 1832-1841.

Petru, E., B. U. Sevin, et al. (1990). "Comparative chemosensitivity profiles in four human ovarian carcinoma cell lines measuring ATP bioluminescence." *Gynecol Oncol* 38(2): 155-160.

Prasad, G., T. Sottero, et al. (2011). "Inhibition of PI3K/mTOR pathways in glioblastoma and implications for combination therapy with temozolomide." *Neuro Oncol* 13(4): 384-392.

Ray, P. (2011). "Multimodality molecular imaging of disease progression in living subjects." *J Biosci* 36(3): 499-504.

Rehemtulla, A., N. Taneja, et al. (2004). "Bioluminescence detection of cells having stabilized p53 in response to a genotoxic event." *Mol Imaging* 3(1): 63-68.

Robbins, D. and Y. Zhao (2011). "Imaging NF-kappaB signaling in mice for screening anticancer drugs." *Methods Mol Biol* 716: 169-177.

Runnels, J. M., A. L. Carlson, et al. (2011). "Optical techniques for tracking multiple myeloma engraftment, growth, and response to therapy." *J Biomed Opt* 16(1): 011006.

Sano, D., F. Matsumoto, et al. (2011). "Vandetanib restores head and neck squamous cell carcinoma cells' sensitivity to cisplatin and radiation in vivo and in vitro." *Clin Cancer Res* 17(7): 1815-1827.

Scabini, M., F. Stellari, et al. (2011). "In vivo imaging of early stage apoptosis by measuring real-time caspase-3/7 activation." *Apoptosis* 16(2): 198-207.

Schuetz, E., L. Lan, et al. (2002). "Development of a real-time in vivo transcription assay: application reveals pregnane X receptor-mediated induction of CYP3A4 by cancer chemotherapeutic agents." *Mol Pharmacol* 62(3): 439-445.

Sevin, B. U., Z. L. Peng, et al. (1988). "Application of an ATP-bioluminescence assay in human tumor chemosensitivity testing." *Gynecol Oncol* 31(1): 191-204.

Shan, L., S. Wang, et al. (2008). "Bioluminescent animal models of human breast cancer for tumor biomass evaluation and metastasis detection." *Ethn Dis* 18(2 Suppl 2): S2-65-69.

Shimomura, O. (2006). *Bioluminesence: Chemical Principles and Methods*, World Scientific Publishing

Spiotto, M. T., A. Banh, et al. (2010). "Imaging the unfolded protein response in primary tumors reveals microenvironments with metabolic variations that predict tumor growth." *Cancer Res* 70(1): 78-88.

Svensson, R. U., J. M. Haverkamp, et al. (2011). "Slow disease progression in a C57BL/6 pten-deficient mouse model of prostate cancer." *Am J Pathol* 179(1): 502-512.

Sweeney, T. J., V. Mailander, et al. (1999). "Visualizing the kinetics of tumor-cell clearance in living animals." *Proc Natl Acad Sci U S A* 96(21): 12044-12049.

Takeshita, F., N. Hokaiwado, et al. (2009). "Local and systemic delivery of siRNAs for oligonucleotide therapy." *Methods Mol Biol* 487: 83-92.

Taneja, P., D. P. Frazier, et al. (2009). "MMTV mouse models and the diagnostic values of MMTV-like sequences in human breast cancer." *Expert Rev Mol Diagn* 9(5): 423-440.
Tang, Y., K. Shah, et al. (2003). "In vivo tracking of neural progenitor cell migration to glioblastomas." *Hum Gene Ther* 14(13): 1247-1254.
Teitz, T., J. J. Stanke, et al. (2011). "Preclinical models for neuroblastoma: establishing a baseline for treatment." *PLoS One* 6(4): e19133.
Tiffen, J. C., C. G. Bailey, et al. (2010). "Luciferase expression and bioluminescence does not affect tumor cell growth in vitro or in vivo." *Mol Cancer* 9: 299.
Tivnan, A., L. Tracey, et al. (2011). "MicroRNA-34a is a potent tumor suppressor molecule in vivo in neuroblastoma." *BMC Cancer* 11: 33.
van der Horst, G., J. J. van Asten, et al. (2011). "Real-time cancer cell tracking by bioluminescence in a preclinical model of human bladder cancer growth and metastasis." *Eur Urol* 60(2): 337-343.
Vikis, H. G., E. N. Jackson, et al. (2010). "Strain-specific susceptibility for pulmonary metastasis of sarcoma 180 cells in inbred mice." *Cancer Res* 70(12): 4859-4867.
Viola, R. J., J. M. Provenzale, et al. (2008). "In vivo bioluminescence imaging monitoring of hypoxia-inducible factor 1alpha, a promoter that protects cells, in response to chemotherapy." *AJR Am J Roentgenol* 191(6): 1779-1784.
Vykhovanets, E. V., S. Shukla, et al. (2008). "Molecular imaging of NF-kappaB in prostate tissue after systemic administration of IL-1 beta." *Prostate* 68(1): 34-41.
Walenta, S., M. Dellian, et al. (1992). "Pixel-to-pixel correlation between images of absolute ATP concentrations and blood flow in tumours." *Br J Cancer* 66(6): 1099-1102.
Walenta, S., T. Schroeder, et al. (2002). "Metabolic mapping with bioluminescence: basic and clinical relevance." *Biomol Eng* 18(6): 249-262.
Wang, H., F. Cao, et al. (2009). "Trafficking mesenchymal stem cell engraftment and differentiation in tumor-bearing mice by bioluminescence imaging." *Stem Cells* 27(7): 1548-1558.
Woolfenden, S., H. Zhu, et al. (2009). "A Cre/LoxP conditional luciferase reporter transgenic mouse for bioluminescence monitoring of tumorigenesis." *Genesis* 47(10): 659-666.
Wu, F., R. Xu, et al. (2008). "In vivo profiling of estrogen receptor/specificity protein-dependent transactivation." *Endocrinology* 149(11): 5696-5705.
Yan, W., D. Xiao, et al. (2011). "Combined bioluminescence and fluorescence imaging visualizing orthotopic lung adenocarcinoma xenograft in vivo." *Acta Biochim Biophys Sin (Shanghai)* 43(8): 595-600.
Zhang, G. J. and W. G. Kaelin, Jr. (2005). "Bioluminescent imaging of ubiquitin ligase activity: measuring Cdk2 activity in vivo through changes in p27 turnover." *Methods Enzymol* 399: 530-549.
Zhang, L., K. C. Lee, et al. (2007). "Molecular imaging of Akt kinase activity." *Nat Med* 13(9): 1114-1119.
Zhang, N., S. Lyons, et al. (2009). "A spontaneous acinar cell carcinoma model for monitoring progression of pancreatic lesions and response to treatment through noninvasive bioluminescence imaging." *Clin Cancer Res* 15(15): 4915-4924.
Zhang, Q., A. A. Triplett, et al. (2010). "Temporally and spatially controlled expression of transgenes in embryonic and adult tissues." *Transgenic Res* 19(3): 499-509.
Zumsteg, A., K. Strittmatter, et al. (2010). "A bioluminescent mouse model of pancreatic {beta}-cell carcinogenesis." *Carcinogenesis* 31(8): 1465-1474.

Part 3

Bacterial Bioluminescence

8

Oscillation in Bacterial Bioluminescence

Satoshi Sasaki
Tokyo University of Technology
Japan

1. Introduction

Live cells show various dynamic characteristics, such as cell division or material production. When we consider that a cell is a chemical reactor that contains an enzyme in its structure, the rates of chemical reaction catalysed by them depend on the cell density. As the amount of enzyme within the cell differs according to the rate of expression of a specific gene, the rate of the reaction also depends on the condition of the cell. In short, chemical reactions caused by cells are nonlinearly related to the cell density; the reaction rate is not proportional to the cell density. This is one remarkable aspect of live cells. In the field of chemical analysis, bacterial cell behaviour is often used. For example, changes in respiration caused by chemical compounds that inhibit the respiratory chain (such as KCN) can be quantified, theoretically, by measuring the changes in the dissolved oxygen concentration.

Biomaterial-based devices have been reported, such as biochips or biosensors. These are not truly "bio" because they use an enzyme or antibody outside of the cell. Microbial sensors (Melidis, P.; Georgiou, D.(2002).; Kang. KH.; Jang. JK.; Pham. TH.; Moon. H.; Chang. IS. & Kim, BH. (2003).; Moon, H.; Chang, IS.; Kang, KH.; Jang, JK. & Kim, BH. (2004). ; Chang, IS.; Moon, H.; Jang, JK. & Kim, BH. (2005).; Kogure, H.; Kawasaki, S.; Nakajima, K.; Sakai, N.; Futase, K.; Inatsu, Y.; Bari, ML.; Isshiki, K. & Kawamoto, S. (2005).; Vaiopoulou, E.; Melidis, P.; Kampragou, E. & Aivasidis, A. (2004).; Yano, Y.; Numata, M.; Hachiya, H.; Ito, S.; Masadome, T.; Ohkubo, S.; Asano, Y. & Imato, T. (2001).; Kim, M.; Hyun, MS.; Gadd, GM.; Kim, GT.; Lee, SJ. & Kim, HJ. (2009). Davila, D.; Esquivel, JP.; Sabate, N. & Mas, J. (2011).), known as the analysis of devices that use live microbial cells as molecular-recognition material, are the only exception. This sensor, however, is based on a shift from one equilibrium to another. For example, a respiration inhibition-based microbial sensor measures a certain toxic compound because the dissolved oxygen concentration near the cells increases when the toxic compound exists. The main reason for the use of microorganisms is that they are more cost-effective than purified enzymes or antibodies. The dynamics of the bacterial cells are not at all used. The nonlinearity of cell behaviour has recently been studied (Wu, BM.; Subbarao, KV. & Qin, QM. (2008).; Kenkre, V. M.; &, Kumar, N. (2008).; Dobrescu, R. & Purcarea, VI. (2011)). A suitable bacterium model is, therefore, needed to start a fundamental study on the nonlinearity of the cell. In our group studies, bioluminescence characteristics have been identified (Sasaki S., Mori Y., Ogawa M., Funatsuka S.,(2010)). Bioluminescent bacteria are those that emit light autonomously without the need of excitation light. The bioluminescence reaction is catalysed by bacterial luciferase (Raushel, F. M. & T. O. Baldwin; (1989), Lee, J., Y. Y. Wang and B. G. Gibson; (1991), Hastings, J. W. (1996), Shirazy, N. H., B. Ranjbar, S. Hosseinkhani, K. Khalifeh, A. R. Madvar and H. Naderi-Manesh (2007)). The reaction requires a flavin mononucleotide (FMNH2), a long-chain aliphatic aldehyde, and O_2 to produce light (Balny, C. and J. W. Hastings (1975), Kurfurst, M., S. Ghisla and J. W. Hastings (1983), Tu, S. C., B. Lei, M. Liu, C. K. Tang and C. Jeffers (2000)).

$$FMNH_2+RCOOH+O_2\text{--->}FMN+RCOOH+H_2O+hv$$

This reaction is catalysed by bacterial luciferase (Karatani, H.; Izuta, T. & Hirayama, S. (2007)). This enzyme is synthesised by a process called quorum sensing, in which the synthesis occurs only after the cells recognise each other to be above a threshold in density. Two substrates, FMNH2 and RCHO (linear alkyl aldehyde), of the reaction are also synthesised in the cell. The substrate with the least amount is, therefore, the rate-determining factor. The intensity of the bioluminescence has been reported primarily in connection with the oxygen concentration, but, theoretically, two other compounds might be candidates. Bacterial luminescence that has been used for environmental monitoring has been reviewed (Girott, S.; Ferri, E.N.; Fumo, M.G.; & Maiolini, E. (2008). Recently, an oscillation in luminescence from a well-stirred bacterial suspension was reported (Sato, Y. and S. Sasaki (2008)). Here, in this chapter, the relationship between the oxygen and oscillation mode was investigated.

Changes in the luminescence spectra are also reported.

2. Experimentals

Bioluminescent bacteria, *Photobacterium kishitanii,* collected from the skin of a cuttlefish and *Todarodes pacificus* (for sashimi), were purified and used. In a well-stirred solution, dissolved oxygen is in equilibrium with the atmospheric oxygen. This may not be the case with a bioluminescent bacterial suspension. As reported above, the luminescent reaction consumes oxygen to produce light. Simultaneously, production of the substrate FMNH2 requires energy that is produced by respiration. Karatani calculated the energy required to produce light and concluded that the bacterial bioluminescence is an extremely oxygen-consuming process. A bioluminescent bacterial suspension was, therefore, suspected to show a very low dissolved oxygen (DO) concentration. In this study, we began with the measurement of both DO and luminescent intensity through the period of oscillation.

As the luminescent reaction occurs inside the cell, the luminescent intensity is affected by the [DO] inside the cell rather than that in the suspension. Because the dynamic measurement of [DO] within a bacterium is considered to be difficult, we focused on any change in cell density during the oscillation period. The colour of bacterial bioluminescence is determined by the fluorescent protein (LumP) (Sato Y, Shimizu S, Ohtaki A, Noguchi K, Miyatake H, Dohmae N, Sasaki S, Odaka M, Yohda M., Crystal structures of the lumazine protein from Photobacterium kishitanii in complexes with the authentic chromophore, 6,7-dimethyl- 8-(1'-D-ribityl) lumazine, and its analogues, riboflavin and flavin mononucleotide, at high resolution., J Bacteriol. 2010 Jan;192(1):127-33.). We then, therefore, measured the spectral change in luminescence through the oscillation period.

2.1 Relationship between the bacterial bioluminescence and dissolved oxygen concentration in a bacterial suspension

Photobacterium belongs to a family of Gram-negative, facultatively aerobic bacteria (Urbanczyk, H.; Ast, JC. & Dunlap, PV. (2011)). We started by measuring the oxygen effect on bioluminescence. The intensity of the bioluminescence was measured using a self-made luminescence detector (five commercially available solar cells were connected in series) or optical power meter (Model 3664, Hioki E.E. Co.). The output voltage generated by both devices was measured and recorded with an A/D converting logger (NR 250, Keyence Co.). An oscillation broth (Yeast extract 2.5 g L-1, Bacto peptone 5 g L-1, and NaCl 30 g L-1) or marine broth (DifcoTM marine broth 2216, Becton, Dickinson, and Company) was prepared and filtrated using a 0.22 µm filter (Nalgene disposable filter unit, Thermo Fisher Scientific, Inc.). A glass cell with an inner diameter of 31 mm was placed over a magnetic stirrer. The schematic illustration of the measurement system is shown in Fig. 1. All the equipment was placed in an incubator (VS401, Versos Co., Ltd.) adjusted at 17°C with 10, 20, 30, and 50 mL of oscillation broth to determine the effects of the air-liquid interface area/volume. In addition, the dilution effect of the marine broth on the oscillation mode was investigated by diluting the broth 1.5 and 3 times. For the simultaneous measurement of luminescence and dissolved oxygen concentration, an optical fibre-based DO sensor (FOXY R, Ocean Optics, Inc.) was placed into the bacterial suspension (Fig. 2).

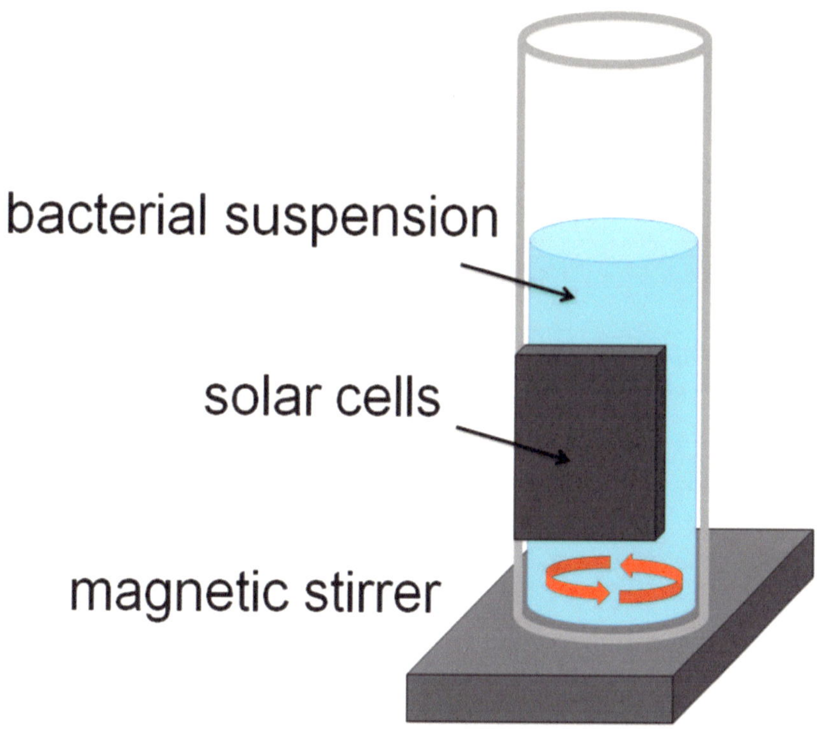

An aluminium foil cap was placed loosely on the glass tube to prevent contamination during the measurement.

Fig. 1. Schematic illustration of the bioluminescence intensity measurement.

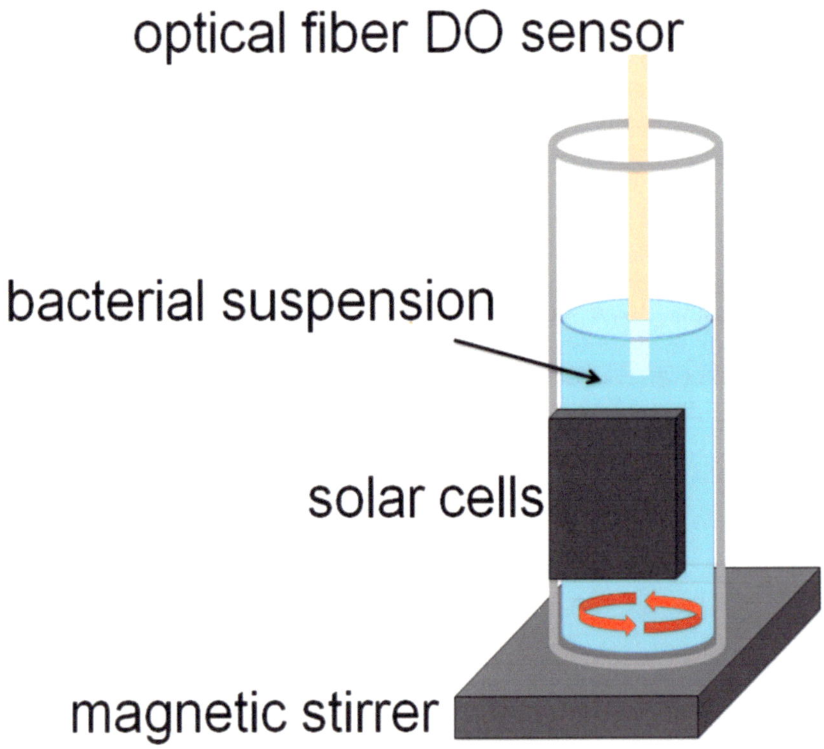

An optical fibre sensor tip was placed vertically in the middle of the bacterial suspension. An aluminium foil cap was placed loosely on the glass tube to prevent contamination during the measurement.

Fig. 2. Schematic illustration of the system for the simultaneous measurement of the luminescence intensity and dissolved oxygen concentration.

2.2 Simultaneous measurement of the luminescence and cell density during oscillation

Continuous measurement of the optical density (OD) of the bacterial suspension was performed using an OD meter (ODBox-A, TAITEC Co.). A 500 mL Erlenmeyer flask with 100 mL of bacterial suspension was set over a rotary shaker (NR-2, TAITEC Co.), and, on the surface of the flask, five solar cells connected in a series were attached (Fig. 3). The generated voltage was measured and recorded by the same logger as reported in 2.1. All the equipment was placed in a self-made dark box, and measurements were performed at room temperature ranging from 20 to 23°C.

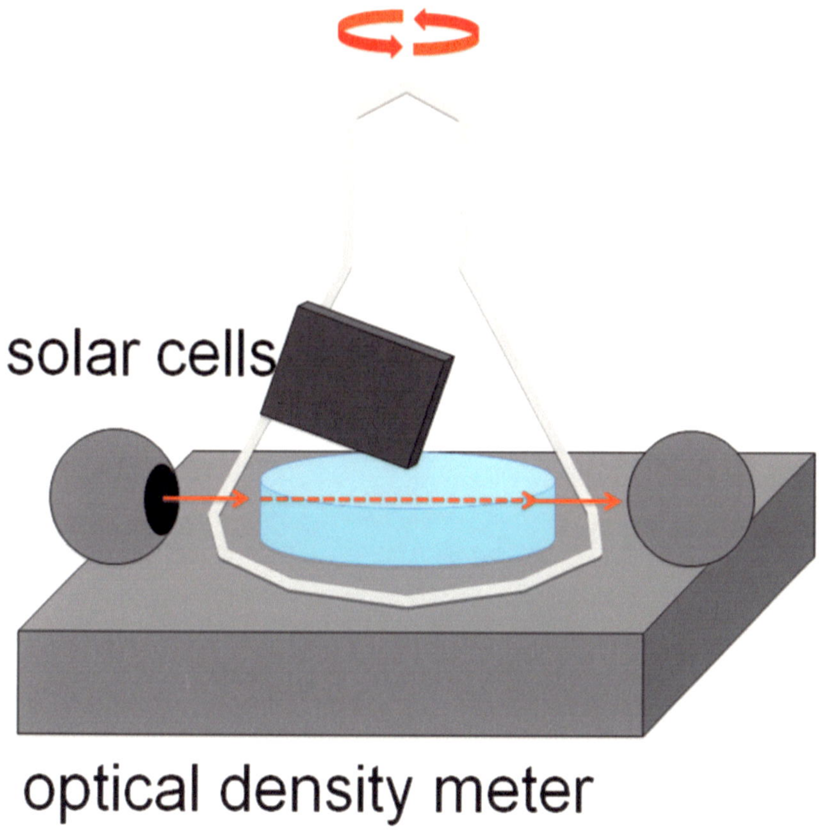

A 500 mL Erlenmeyer flask with 100 mL oscillation broth was shaken at 100 rpm. Solar cells were attached on the flask surface. All the optical setup was enclosed in a self-made dark box.

Fig. 3. Experimental setup for the simultaneous measurement of the luminescence and cell density.

2.3 Spectral change in the bacterial bioluminescence during oscillation

Two optical filters that transmit wavelengths of 479 and 521 nm (Optical Coatings Japan) were placed to cover the sensor windows of the optical power meter (Fig. 3).

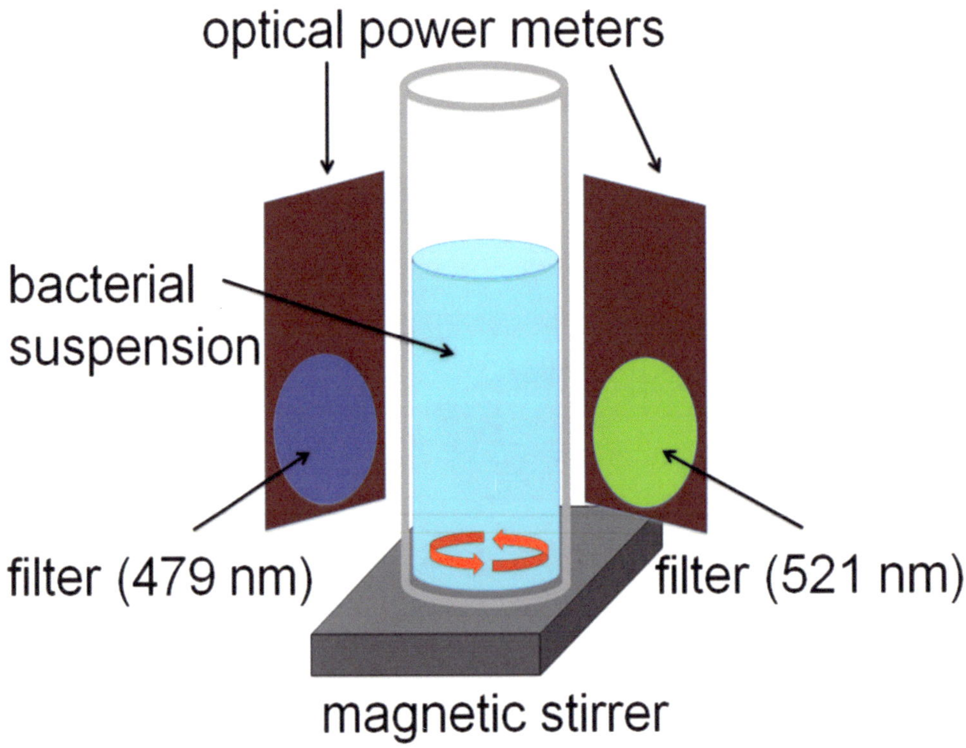

Two sensors with filters were placed at the same height so that the stirrer bar did not affect the measurement. All the optical setup, including the magnetic stirrer, was enclosed in the incubator.

Fig. 4. System setup for the measurement of spectral change.

3. Results and discussion

The effects of the suspension volume on the oscillation mode are shown in Fig. 5. Remarkable oscillatory waves were observed in the case of 10, 20, and 30 mL but not in the case of 50 mL. It was noteworthy that, even with the largest volume, the 50 mL suspension showed the smallest luminescent intensity. This might be due to the shortage of the oxygen supply, as the fixed liquid-air interface area could allow a fixed amount of oxygen diffusion into the suspension. In the case of 50 mL, the distributed oxygen to each cell should be smaller than in the case of other volumes. As the case of 30 mL showed the most distinct oscillation, this volume was chosen for further experiments.

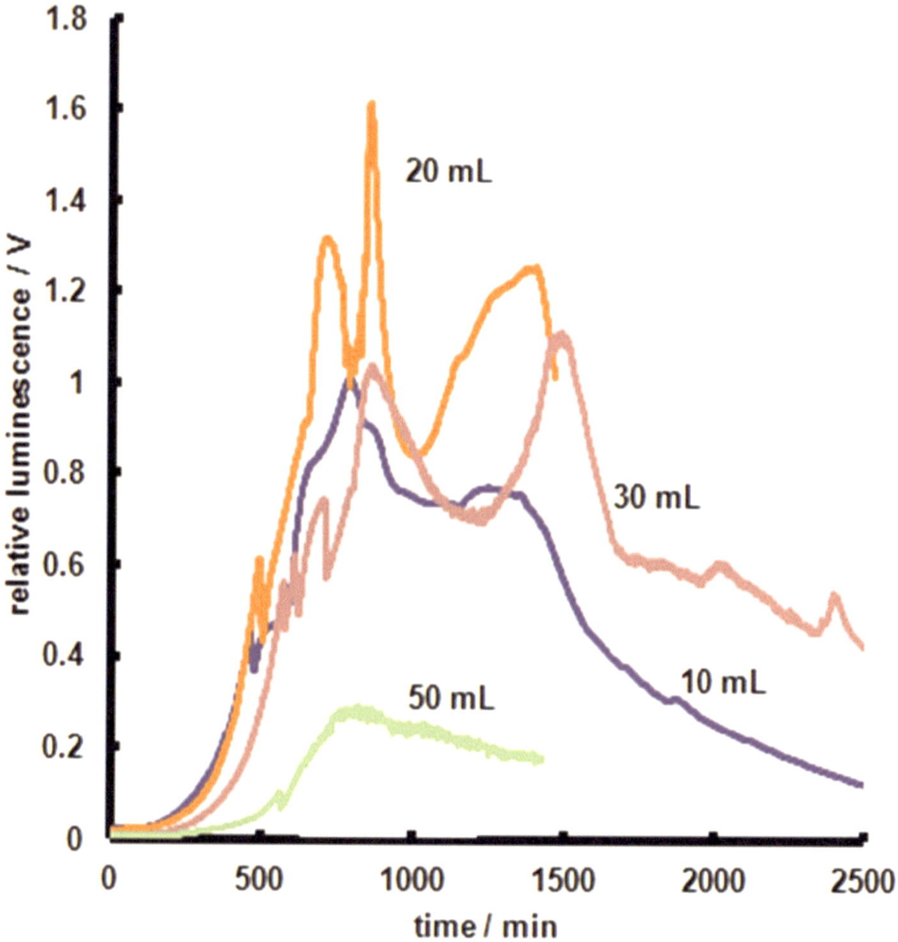

P. kishitanii was inoculated into oscillation broth with each volume in a 22 mm diameter glass tube. The suspension was stirred using a magnetic stirrer. The temperature was maintained at 17°C.

Fig. 5. Suspension volume effect on the oscillation in bacterial bioluminescence.

First, oscillation in bioluminescence was observed only in the case with the oscillation broth (Sato, Y. & S. Sasaki (2008)). We were interested in the use of the common marine broth and tried to determine the broth dilution effect on the mode of oscillation (whether or not it oscillated) because, in our previous report, the oscillation was thought to be the result of a lack of nutrients. Therefore, even with the marine broth, the oscillation was observed (Fig. 6). In cases of no dilution, clearer peaks were observed than in the cases with dilution. In addition, the luminescence measurement was performed with a cap on the glass tube. This case also showed, though with a different mode, an oscillatory behaviour. Through measurement with two different broths, the effect of oxygen supply into the suspension on the mode change of oscillation was strongly indicated.

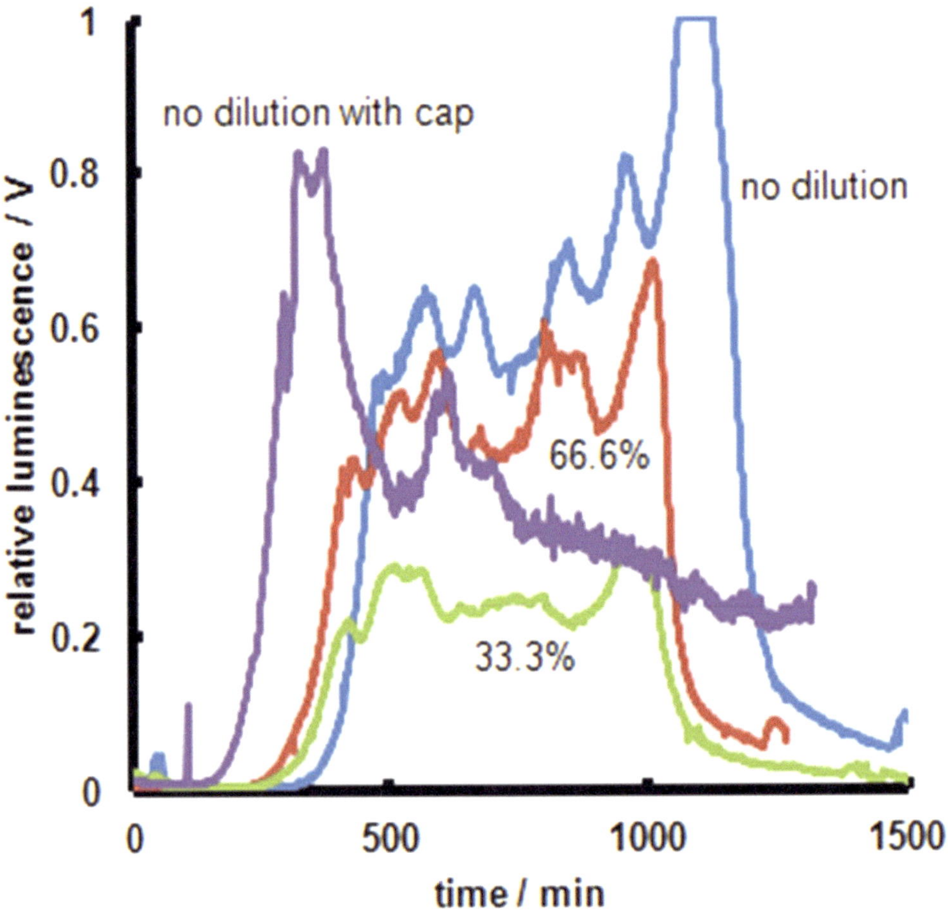

The measurement conditions were the same as those reported in Fig. 5.

Fig. 6. Broth-dilution effect on the oscillation in bacterial bioluminescence.

Encouraged by the above results, we tried to see the stirring effect on the bioluminescence over a shorter period because, during measurements lasting more than a day (1,440 min), the cell density effect on the luminescence could not be ignored. We, therefore, used brightly glowing suspensions (5 - 19 hours after inoculation / 10^8-10^9 cells mL^{-1}) and investigated the effect of stirring on the luminescence intensity. First, the dark suspension was stirred until the luminescence reached a stable intensity. The result is shown in Fig. 7 (a). The luminescence intensity was gradually increased. This might be due to the increase in the fluorescence activity of LumP. In other bioluminescent bacteria, *V. fischeri* Y1, a fluorescent protein changes the fluorescent activity in its redox states; i.e., when reduced, the fluorescence is lost, and, when oxidised, the original fluorescence is retrieved (Karatani, H.; Izuta, T.; & Hirayama, S. (2007)). LumP in P. kishitanii might have similar characteristics.

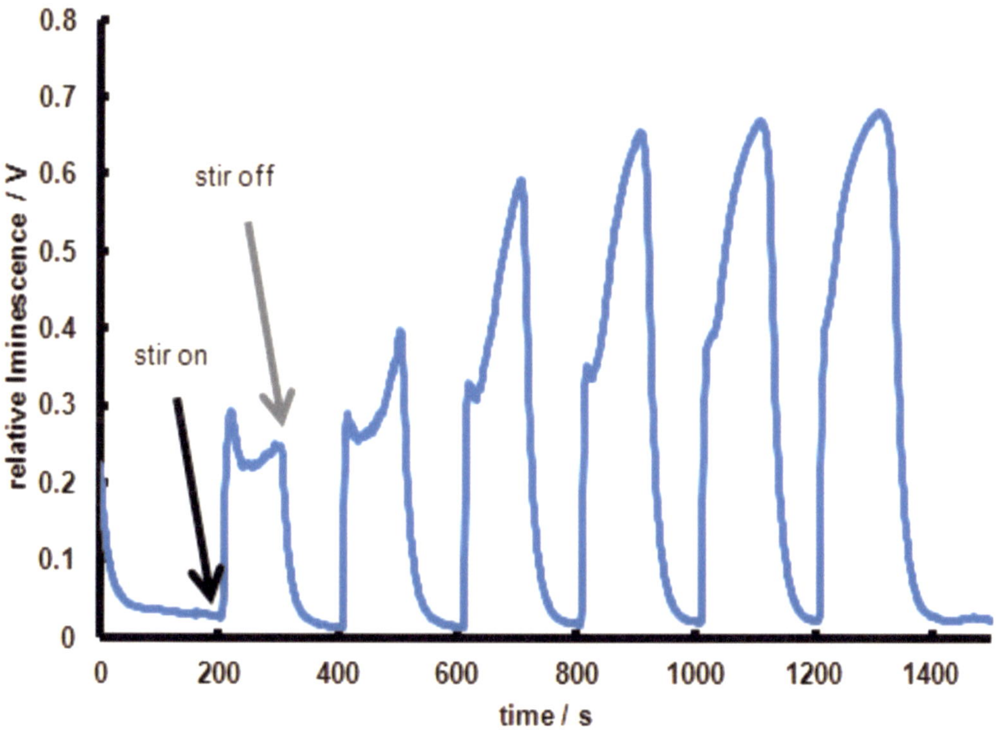
0.8
0.7
0.6
0.5
0.4
0.3
0.2
0.1
0
relative lminescence / V
stir off
stir on
0
200
400
600
800
1000
1200
1400
time / s

(a)

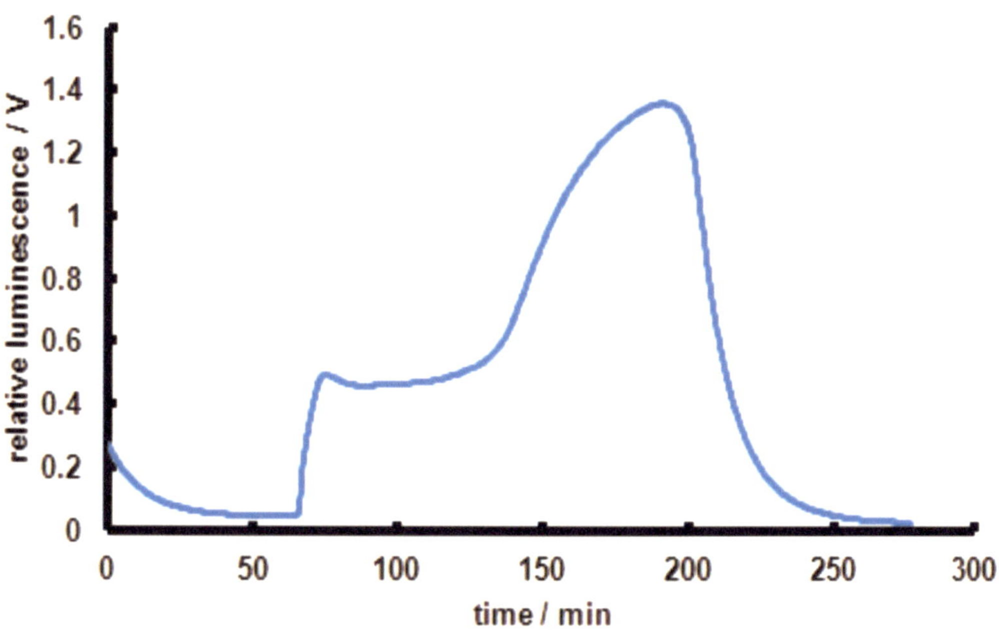
1.6
1.4
1.2
1
0.8
0.6
0.4
0.2
0
relative luminescence / V
0
50
100
150
200
250
300
time / min

(b)

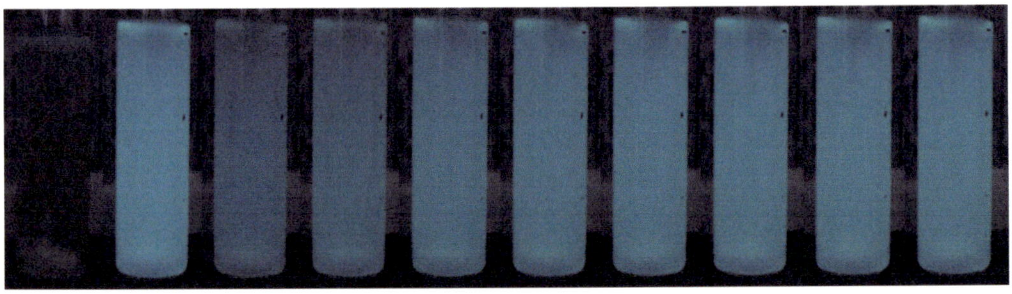

(c)

In (a), the stirrer was switched on at 200, 400, 600, 800, 1,000, and 1,200 min and off at 300, 500, 700, 900, and 1,100 min. In (b), the stirrer was switched on at 60 s and off at 180 s. The measurements in both (a) and (b) were performed at 17°C. Photographs in (c) were taken at a 5 s interval.
Luminescence from the suspension after the stirrer was switched on was measured for two minutes (Fig. 7 (b)). A local maximal luminescence was observed right after the stirring (ca. 60 s), and then, a gradual increase was observed. This characteristic might be related to the LumP fluorescence ability, but the photographs of the luminescence showed no significant colour change (Fig. 7 (c)).

Fig. 7. Time course of the luminescence from the dark suspension after repeated stirring (a), a typical luminescence curve showing two peaks of intensity (b), and interval photographs of luminescence from the suspension in experiment (b) (c).

The effect of stirring on the bright (originally well-stirred) suspension luminescence resulted in different outcomes (Fig. 8). The luminescence increased after switch-off and decreased after switch-on. This tendency is the opposite of the results in Fig. 7 (a). The reason for the decreasing tendency of luminescence under the stirred condition is difficult to explain as long as we regard the suspension to be homogeneous. As is reported later, the condition of the cells in the suspension seemed to be inhomogeneous.

The suspension DO characteristic during the oscillation is shown in Fig. 9. As is evident from the figure, the DO during the oscillation was approximately zero. This result was considered to be reasonable, since the origin of bioluminescence was an oxygen-quenching mechanism. One evolutionary purpose of bioluminescence is oxygen quenching (Rees, J.F (1998), Timmins, GS. (2001), Szpilewska, H., Czyz, A. & Wegrzyn, G. (2003)). In a well-stirred condition, oxygen in the atmosphere diffused into the suspension, but most of it was assumed to be consumed by both the luminescence reaction and respiration. *Vibrio fisheri* was reported to perform anaerobic respiration using a certain gene regulator (Septer, AN.; Bose, JL.; Dunn, AK. & Stabb, EV. (2010).). No such report was available for the *Photobacterium* species. As a result, there was no significant relationship between the suspension DO and oscillatory waves. From this result, we recognised the importance of considering the DO within rather than outside the cell.

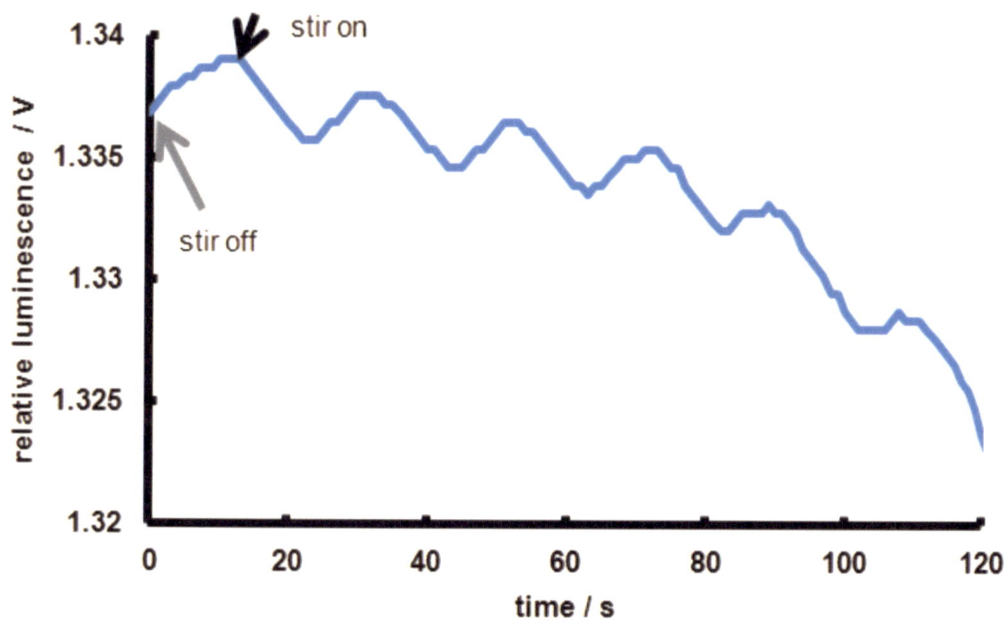

The stirrer was switched off at 0, 20, 40, 60, 80, and 100 s and on at 10, 30, 50, 70, 90, and 110 s. The measurement was performed at 17°C.

Fig. 8. Effect of stirring on the bright suspension.

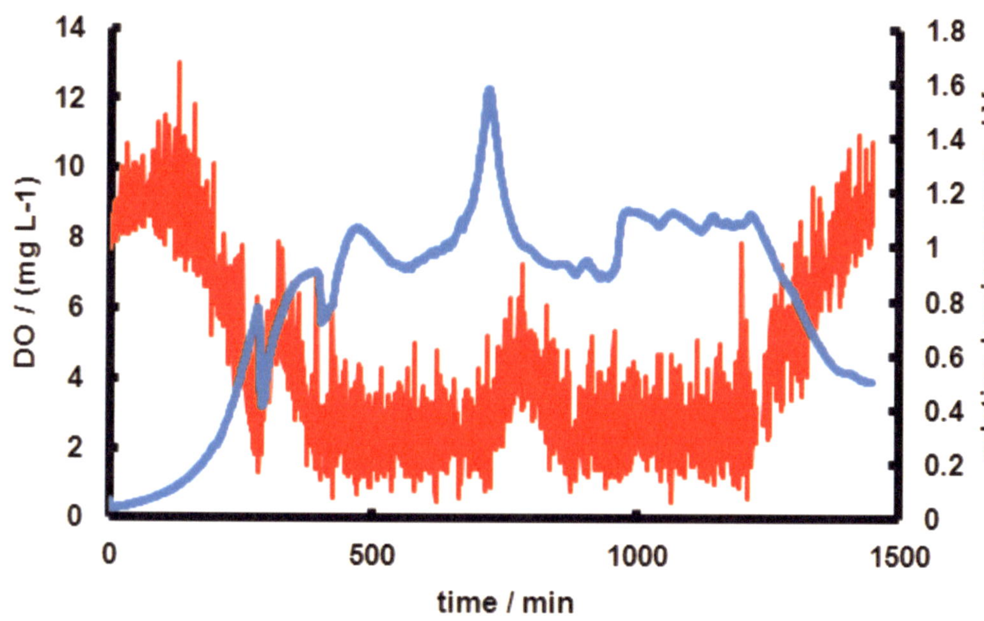

Data was recorded every ten minutes.

Fig. 9. Time courses of dissolved oxygen and luminescence.

The cell density was expressed by the optical density (OD) in the measurement. OD was measured as the decrease in near-infrared light measured at the sensor (Fig. 3). This OD probe light did not affect the bioluminescence measurement using solar cells. Four results of the simultaneous measurement of DO and luminescence are shown in Fig. 10 (a) - (d). We searched for the common characteristics between the DO and luminescent curves in the four cases and found that, after the luminescence peak, a plateau in the DO curve appeared. This might be due to the decrease in DO inside the cell after the luminescence that inhibited the respiration. Lack of oxygen might have suppressed the energy production by the respiration.

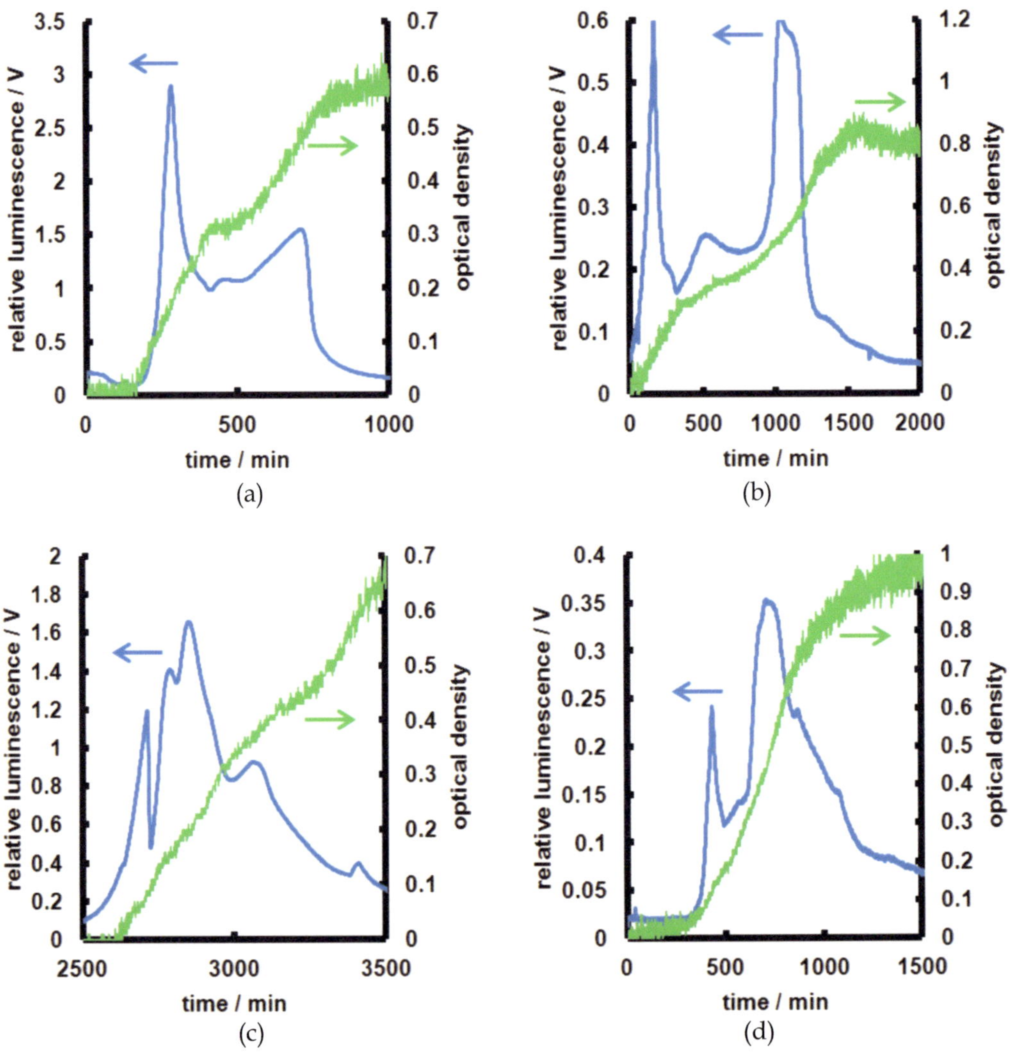

A 100 mL oscillation broth in a 500 mL Erlenmeyer flask was used for each measurement.
Measurements were performed at room temperature (20-23°C).

Fig. 10. Time courses of the luminescence and optical density in four experiments under the same condition.

The oscillation mode observed under the same suspension condition differed, as shown in the figures. These differences should be kept in mind for the following experiments. As reported above, the luminescence from LumP (peak wavelength: ca. 475 nm) was the main part of the observed light. The ratio of the luminescence at throughout the oscillation was estimated by the use of optical filters. The results are shown in Fig. 11 (a). A blue light with a spectral peak at 479 nm appeared ca. 1 h after a green light (521 nm) and quenched 4 h before that. This result indicated the change in the fluorescence ability at the beginning and at the end of the oscillation. When the luminescence intensity at 521 nm was plotted against that at 479 nm, the two showed a linear relationship (Fig. 11 (b)). This indicated that the LumP fluorescence ability was stable during the oscillation period.

For the first time, we found an oscillation in bioluminescence intensity. The next step would be to identify the initial reason for the oscillation. Since a definitive answer is not yet available, we propose the hypothesis explained below. Bacterial luminescence spectral change has been reported (Eckstein, JW.; Cho, KW.; Colepicolo, P.; Ghisla, S.; Hastings, JW. & Wilson, T. (1990).; Karatani, H.; Matsumoto, S.; Miyata, K.; Yoshizawa, S.; Suhama, Y. & Hirayama, S. (2006).; Karatani, H.; Yoshizawa, S. & Hirayama, S. (2004).). Under the DO-rich condition, the LumP fluorescence capacity is high, and a blue light is evident, whereas, under a DO-poor condition, luciferin-luciferase luminescence (with a peak wavelength of 540 nm) occupies the main part, and a green light is evident. When the luminescence spectra measured with and without stirring were compared, a slight difference in the peak wavelength was observed (Fig. 12). This result agreed with the above-mentioned report.

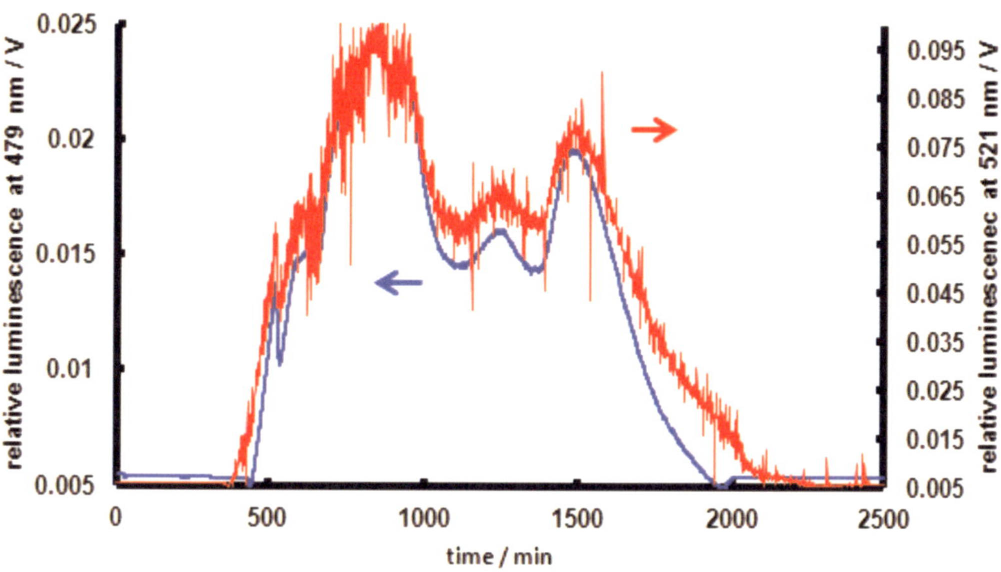

(a)

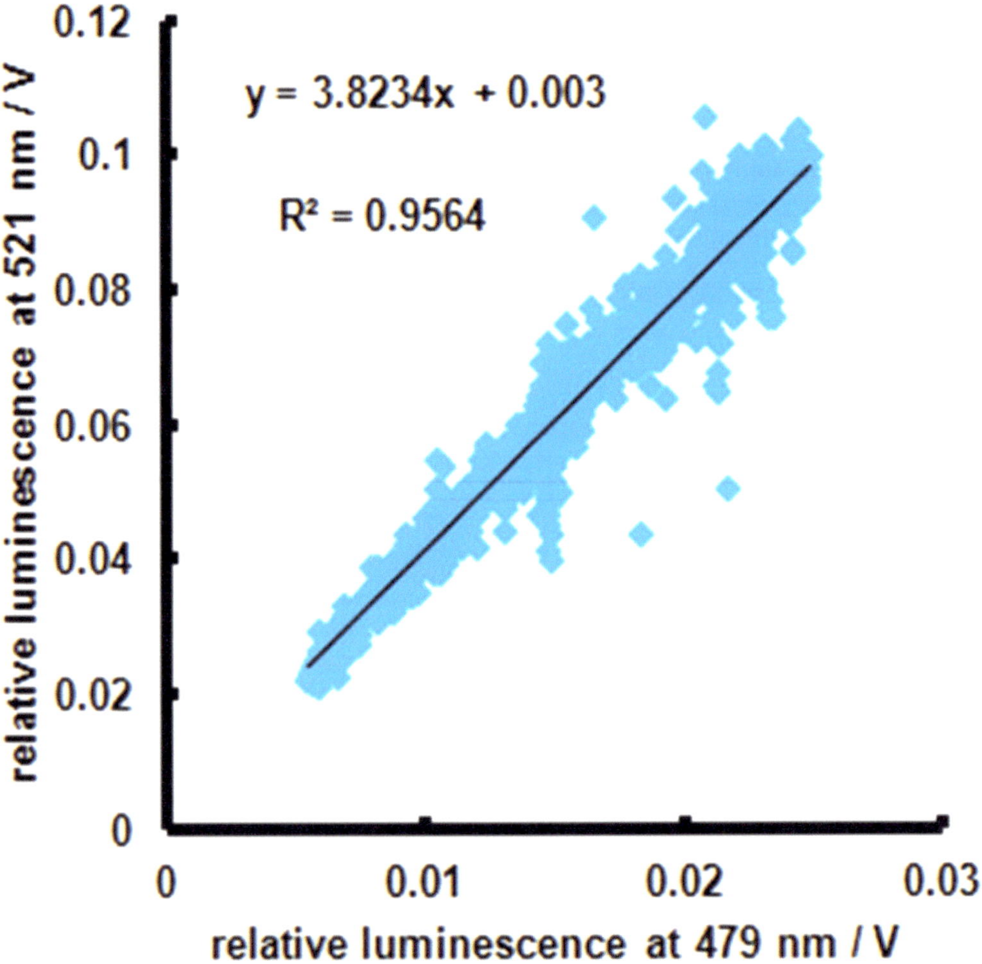

(b)

An approximation line between the two luminescences is illustrated. The coefficient of determination (R^2) was calculated to be 0.9564.

Fig. 11. Bioluminescence oscillation observed in two colours (a) and relationship between blue (479 nm) and green (521 nm) colours (b).

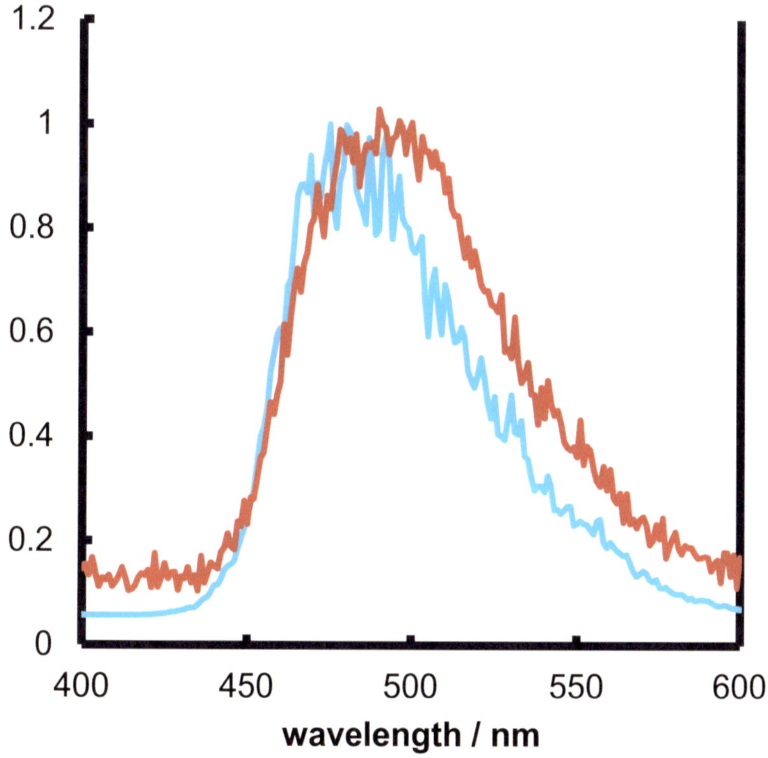

The blue curve indicates the spectrum of luminescence at 479 nm, and the brown curve indicates that at 521 nm.

Fig. 12. Bioluminescence spectra with and without stirring (normalized).

During cell cultivation, the variety of cell phases was assumed to increase with cell growth even when the inoculated cells had the same, synchronised cell phases. In the glowing suspension, the cell condition was assumed to be inhomogeneous. A photograph of the bioluminescent suspension after the stirrer was switched off is shown in Fig. 13. A slowly precipitating block of cells was glowing as brightly as the air-liquid interface part. At that moment, the DO in the middle of the suspension was zero. Unlike others, this block of cells emitted light even under the [DO]=0 condition.

The image was photographed using a digital still camera (GR Digital 3, Ricoh Company, Ltd.) with exposure time of 1/20 s, ISO 1600, f/1.9. The raw image was modified to enhance the contrast using image software (ImageJ).

Fig. 13. Image of brightly glowing cell block precipitating in the suspension.

The results in Fig. 10 indicated the possibility that the luminescence affected the cell growth; i.e., an increase in luminescence caused oxygen deficiency and inhibited the respiration needed for cell growth. Cell growth was assumed to be expressed by the time derivative of the optical density. We, therefore, plotted the time courses of relative luminescence and the time derivative of OD in the same time scale (Fig. 14 (a)). The result shown in Fig. 10 (c) was used because it showed five obvious peaks in the relative luminescence curve. As is clear in Fig. 14 (a), the peaks and valleys in the luminescence curve coincided with those in the time derivative of the optical density. We then plotted the derivative against the relative luminescence (Fig. 14 (b)). The obtained curve showed that the two parameters were in the relationship with a negative Pearson product-moment correlation coefficient.

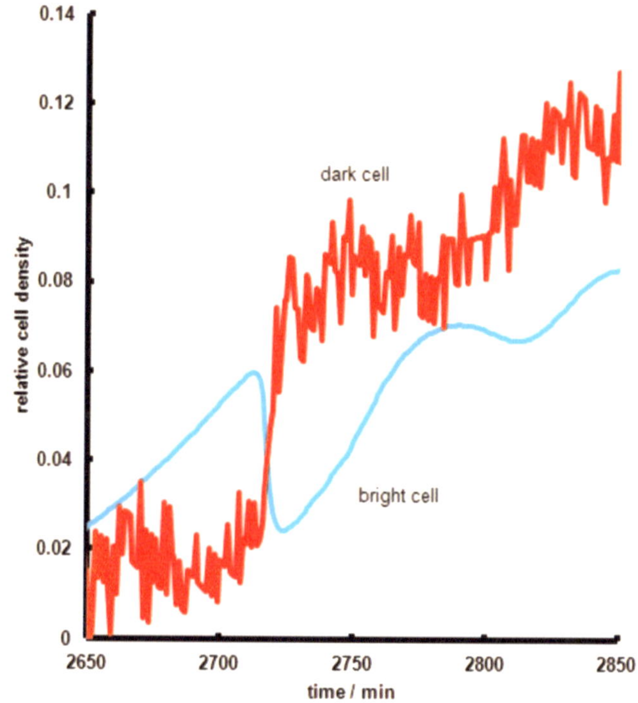

(a)

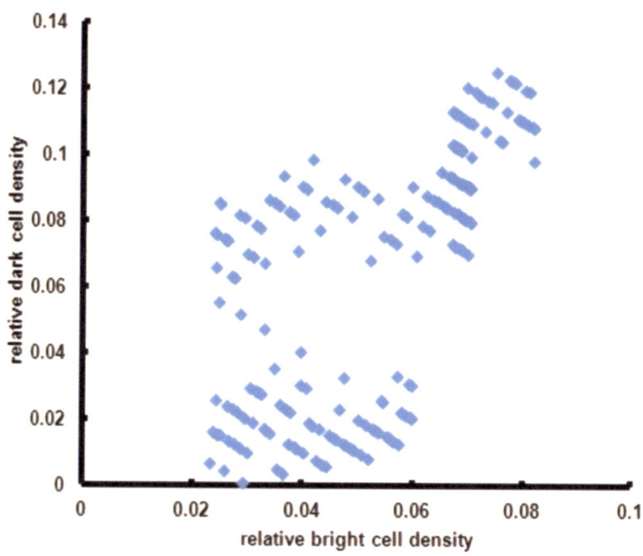

(b)

In (a), the relative bright cell density was calculated as 0.05* (relative luminescence), whereas the relative dark cell density was calculated as {OD-0.05*(relative luminescence)}. In (b), data at 2650 - 2850 min were chosen.

Fig. 14. Time courses of bright and dark cells (a) and relative dark cell density plotted against the relative bright cell density (b).

The type of model that could describe such oscillatory behaviour should be identified. One of the best-known models is the one proposed by Alfred Lotka and, later, by Vito Volterra (Mounier, J.; Monnet, C.; Vallaeys, T.; Arditi, R.; Sarthou, AS.; Helias, A. & Irlinger, F. (2008).; Varon, M. & Zeigler, BP. (1978).; Tsuchiya, HM.; Drake, JF.; Jost, JL. & Fredrickson, AG. (1972).) This model is often used to characterise predator-prey interactions. If we were to adjust the bacterial bioluminescence in the model, the following might be examples:

$$\begin{aligned} &broth + bright\ cell \rightarrow 2\,bright\ cell \\ &bright\ cell + dark\,cell \rightarrow 2\,dark\,cell \\ &dark\,cell \rightarrow dead\,cell \end{aligned} \tag{1}$$

In these reactions, we regarded that

1. one bright cell divides into two bright cells with the supply of infinite broth;
2. one bright cell becomes a dark cell as a result of interaction with a dark cell (both cells consume oxygen as a result of respiration and become dark ones);
3. a dark cell becomes a dead cell.

If we write

A: broth, X: bright cell, Y: dark cell, P: dead cell, then the above equations can be written as

$$\begin{aligned} &A + X \xrightarrow{k_1} 2X \\ &X + Y \xrightarrow{k_2} 2Y \\ &Y \xrightarrow{k_d} P \end{aligned} \tag{2}$$

We consider A, the broth, to be infinite and not to decrease through the oscillation reaction (however, in an experiment, it does). As X and Y are the function of the time t, we can write two equations, such as,

$$\begin{aligned} \frac{dX}{dt} &= k_1[A][X] - k_2[X][Y] \\ \frac{dY}{dt} &= k_2[X][Y] - k_d[Y] \end{aligned} \tag{3}$$

These are the typical equations that appear in the model. We have a numerical solution of the two equations, i.e., the time course of X and Y through the simulation using a common spreadsheet software that runs on a personal computer.

Model (1) is not proved to interpret what is going on in the oscillation, but we can approach the real image of the oscillatory reaction. By changing the parameters k_1, k_2, and k_d, we will have curves that look like what we observe, and we should then determine the values for the three parameters and evaluate their suitability from a biochemical viewpoint.

As reported in relation to Fig. 13, luminescence from the suspension with a volume of several tens - hundreds of mL might contain luminescence from cells of different conditions. Future investigation of cells with similar conditions is indicated, therefore, to be necessary. The relationship between the bacterial motility and luminescence was investigated (Sasaki,

S.; Okamoto, T. & Fujii T. (2009)). The evaluation of surface-adsorbed cells was thought to be an effective way for this purpose. The characteristics of the luminescence from ca. 1.0 X 10^6 cells adsorbed on a glass surface are shown in Fig. 16. Irradiation of the cells was performed using a near-UV light (UV lamp—long wavelength, # 166-0500EDU, BIO RAD). The irradiation has the potential to cause a change in the redox state of FMN or other materials that produce an increase in luminescence. Bacterial bioluminescence from the electromagnetic viewpoint has been studied (Pooley DT. (2011)). Investigation of this luminescence from physico-chemical as well as biochemical viewpoints would be needed to explain the entire image of bacterial bioluminescence.

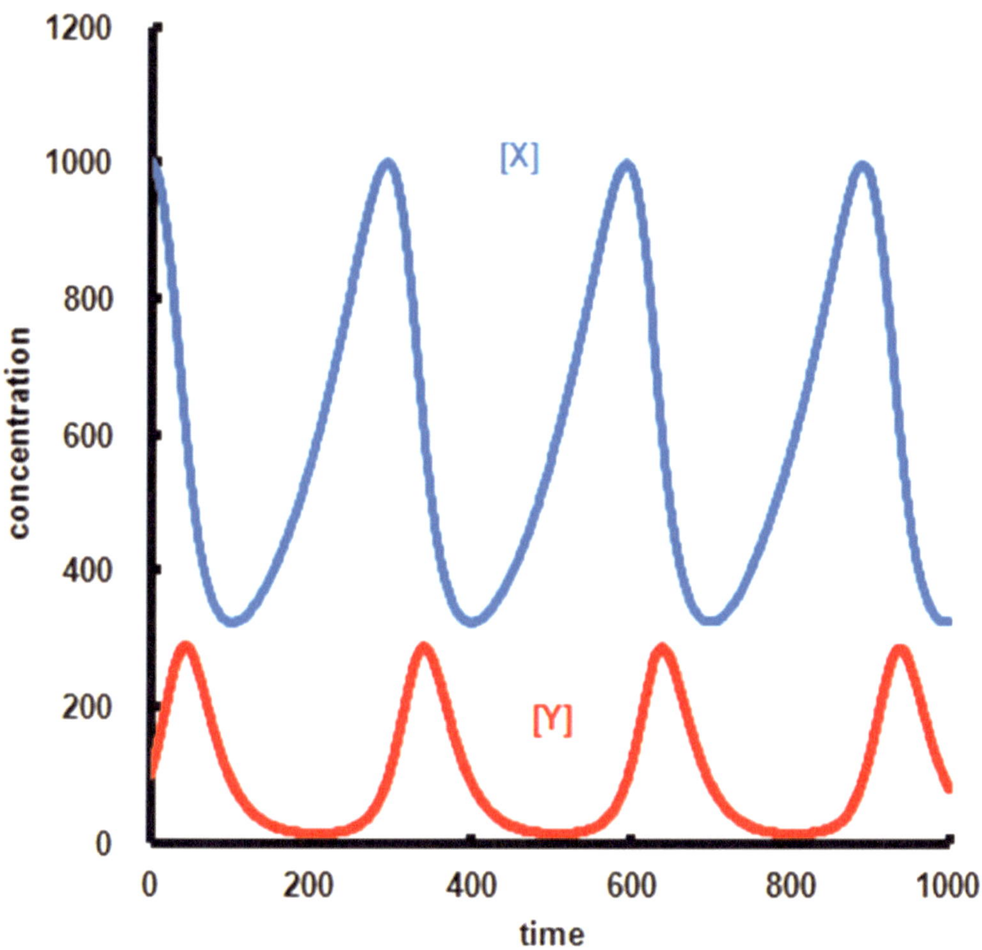

The initial values were [X[=1,000 and [Y]=100, with constants k_1=0.009, k_2=0.06, and k_d=0.0001. The integration time was set at 1, and the calculation was performed using Microsoft Excel 2007 running on a personal computer.

Fig. 15. Solution of Equation (2) using a numerical calculation (Runge-Kutta method)

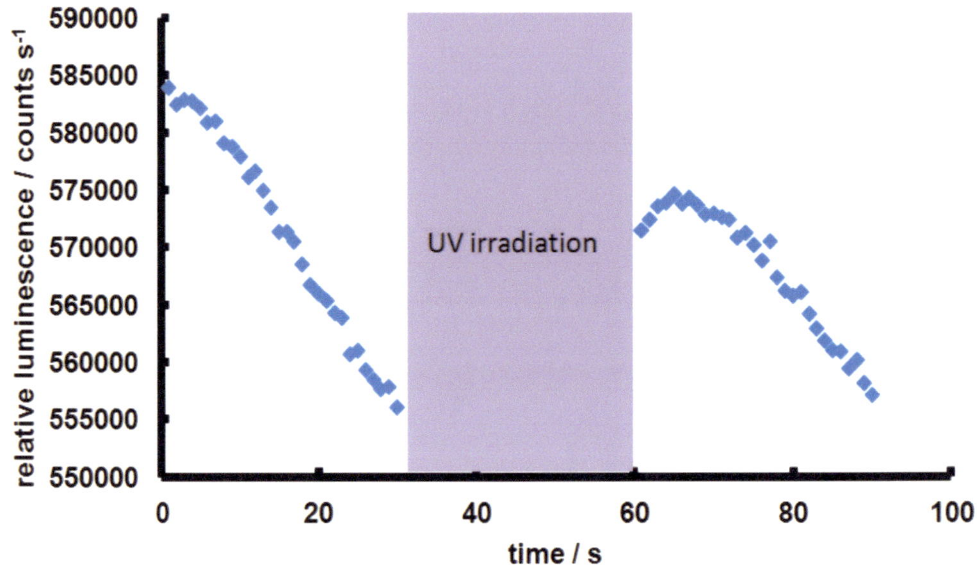

Oil was used to prevent the bacterial environment from drying. Glass with an amino group modification (MAS coated glass slides, Matsunami Glass Ind., Ltd.) was used for the adsorption. The glass was soaked in a marine-broth-based bacterial suspension overnight. A measurement was performed using a luminescence meter (GENE LIGHT GL-200S, Microtec Nichion).

Fig. 16. Effect of irradiation to the luminescence from cells adsorbed on a glass surface.

4. Conclusion

Oscillation in the bacterial bioluminescence mode is strongly dependent on the amount of oxygen supply to the solution. There is no clear relationship between the DO concentration and luminescence intensity, perhaps due to the consumption of oxygen by both the luminescence and respiration. The oscillation occurred at a very low DO concentration, and, when the time course of cell density was plotted with the same timescale as the luminescence intensity, the cell growth rate seemed to decrease after the strong luminescence. The fluorescence ability of LumP seemed constant during the oscillation period, but, at the beginning and at the end, it seemed to decrease. The characterisation of luminescence from a smaller number of cells would be necessary for further investigation of oscillation, considering that the suspension is a mixture of cell groups with a variety of cell phases.

5. Acknowledgments

The author thanks Dr. Hajime Karatani of the Kyoto Institute of Technology for his participation in discussions and Shoji Yamada, Kenshin Tamura, Shingo Kuriyama, Mika Mochizuki, and Hajime Kimoto for their assistance with the experiments.

6. References

Aivasidis, A.; Melidis, P.; Georgiou, D. (2002). Use of a microbial sensor: a new approach to the measurement of inhibitory effects on the microbial activity of activated sludge. *Bioprocess and Biosystems Engineering*, Vol. 25, No. 1, (April 2002), pp. 29-33, ISSN 1615-7591

Balny, C. & Hastings, J. W. (1975). Fluorescence and Bioluminescence of Bacterial Luciferase Intermediates. *Biochemistry* , Vol. 14, No. 21, (October 1975), pp. 4719-4723, ISSN 0006-2960

Chang, IS.; Moon, H.; Jang, JK. & Kim, BH. (2005). Improvement of a microbial fuel cell performance as a BOD sensor using respiratory inhibitors. *Biosensors and Bioelectronics* , Vol. 20, No. 9, (March 2005), pp. 1856-1859, ISSN 0956-5663

Davila, D.; Esquivel, JP.; Sabate, N. & Mas, J. (2011). Silicon-based microfabricated microbial fuel cell toxicity sensor. *Biosensors and Bioelectronics* , Vol. 26, No. 5, (October 2010), pp. 2426-2430, ISSN 0956-5663

Dobrescu, R. & Purcarea, VI. (2011). Emergence, self-organization and morphogenesis in biological structures. *Journal of medicine and life*, Vol. 4, No. 1, (February 2011), pp. 82-90, ISSN 1844-122x

Eckstein, JW.; Cho, KW.; Colepicolo, P.; Ghisla, S.; Hastings, JW. & Wilson, T. (1990). A time-dependent bacterial bioluminescence emission spectrum in an in vitro single turnover system: energy transfer alone cannot account for the yellow emission of *Vibrio fischeri* Y-1. *Proceedings of the National Academy of Sciences U S A*, Vol. 87, No. 4, (February 1990), pp. 1466-1470, ISSN 0027-8424

Girott, S.; Ferri, E. N.; Fumo, M. G.; & Maiolini, E. (2008). Monitoring of environmental pollutants by bioluminescent bacteria. *Analytica Chimica Acta* , Vol. 608, No. 1, (February 2008), pp. 2-29, ISSN 0003-2670

Hastings, J. W. (1996). Chemistries and Colors of Bioluminescent Reactions: a Review. *Gene*, Vol. 173, No. 1, (July 1995), pp. 5-11, ISSN 0378-1119

Kang. KH.; Jang. JK.; Pham. TH.; Moon. H.; Chang. IS. & Kim, BH. (2003). A microbial fuel cell with improved cathode reaction as a low biochemical oxygen demand sensor. *Biotechnology Letters* , Vol. 25, No. 16, (August 2003), pp. 1357-1361, ISSN 0141-5492

Karatani, H.; Izuta, T. & Hirayama, S. (2007). Relationship between the redox change in yellow fluorescent protein of *Vibrio fischeri* strain Y1 and the reversible change in color of bioluminescence *in vitro*. *Photochemical and Photobiological Sciences*, Vol. 6, No. 5, (January 2007), pp. 566-570, ISSN 1474-905X

Karatani, H.; Matsumoto, S.; Miyata, K.; Yoshizawa, S.; Suhama, Y. & Hirayama, S. (2006). Bioluminescence color modulation of *Vibrio fischeri* strain Y1 coupled with alterable levels of endogenous yellow fluorescent protein and its fluorescence imaging.

Photochemistry and Photobiology, Vol. 82, No. 2, (March 2006), pp. 587-592, ISSN 0031-8655

Karatani, H.; Yoshizawa, S. & Hirayama, S. (2004). Oxygen triggering reversible modulation of Vibrio fischeri strain Y1 bioluminescence *in vivo*. *Photochemistry and Photobiology*, Vol. 79, No. 1, (January 2004), pp. 120-125, ISSN 0031-8655

Kenkre, V. M.; &, Kumar, N. (2008). Nonlinearity in bacterial population dynamics: Proposal for experiments for the observation of abrupt transitions in patches. *Proceedings of the National Academy of Sciences U S A*, Vol. 105, No. 48, (November 2008), pp. 18752-18757, ISSN 0027-8424

Kim, M.; Hyun, MS.; Gadd, GM.; Kim, GT.; Lee, SJ. & Kim, HJ. (2009). Membrane-electrode assembly enhances performance of a microbial fuel cell type biological oxygen demand sensor. *Environmental Technology*, Vol. 30, No. 4, (April 2009), pp. 329-336, ISSN 0959-3330

Kogure, H.; Kawasaki, S.; Nakajima, K.; Sakai, N.; Futase, K.; Inatsu, Y.; Bari, ML.; Isshiki, K. & Kawamoto, S. (2005). Development of a novel microbial sensor with baker's yeast cells for monitoring temperature control during cold food chain. *Journal of Food Protection* , Vol. 68, No. 1, (January 2005), pp. 182-186, ISSN 0362-028X

Kurfurst, M.; Ghisla, S. & Hastings, J. W. (1983). Bioluminescence Emission from the Reaction of Luciferase-Flavin Mononucleotide Radical with O2- • *Biochemistry* , Vol. 22, No. 7, (March 1983), pp. 1521-1525, ISSN 0006-2960

Lee, J.; Wang, Y. Y. & Gibson, B. G. (1991). Electronic Excitation Transfer in the Complex of Lumazine Protein with Bacterial Bioluminescence Intermediates. *Biochemistry* , Vol. 30, No. 28, (July 1991), pp. 6825-6835 ISSN 0006-2960

Moon, H.; Chang, IS.; Kang, KH.; Jang, JK. & Kim, BH. (2004). Improving the dynamic response of a mediator-less microbial fuel cell as a biochemical oxygen demand (BOD) sensor. *Biotechnology Letters* , Vol. 26, No. 22, (November 2004), pp. 1717-1721, ISSN 0141-5492

Mounier, J.; Monnet, C.; Vallaeys, T.; Arditi, R.; Sarthou, AS.; Helias, A. & Irlinger, F. (2008). Microbial interactions within a cheese microbial community. *Applied and Environmental Microbiology*, Vol. 74, No. 1, (November 2007), pp. 172-181, ISSN 0099-2240

Pooley DT. (2011). Bacterial bioluminescence, bioelectromagnetics and function. *Photochemistry and Photobiology*, Vol. 87, No. 2, (March 2011), pp. 324-328, ISSN 0031-8655

Raushel, F. M. & Baldwin, T. O. (1989). Proposed Mechanism for the Bacterial Bioluminescence Reaction Involving a Dioxirane Intermediate. *Biochemical and Biophysical Research Communications*, Vol. 164, No. 3, (November 1989), pp. 1137-1142 ISSN 0006-291X

Rees, J. F., B. de Wergifosse, O. Noiset, M. Dubuisson, B. Janssens and E. M. Thompson; (1998). The Origins of Marine Bioluminescence: Turning Oxygen Defense Mechanisms into Deep-Sea Communication Tools, *The Journal of Experimental Biology*, Vol. 201, No. 8, (April 1998), pp. 1211-1221, ISSN 1010-061X

Sasaki, S.; Okamoto, T. & Fujii, T., (2009). Bioluminescence intensity difference observed in luminous bacteria groups with different motility. *Letters in Applied Microbiology*, Vol. 48, No. 3, (March 2009), pp. 313-317, ISSN 0266-8254

Sasaki, S.; Mori, Y.; Ogawa, M. & Funatsuka, S. (2010). Spatio-temporal control of bacterial suspension luminescence using a PDMS cell, *Journal of Chemical Engineering of Japan*, Vol. 43, No. 11, (August 2010), pp. 960-965, ISSN 0021-9592

Sato, Y. & Sasaki, S. (2006). Control of the Bioluminescence Starting Time by Inoculated Cell Density. *Analytical Sciences*, Vol. 22, No. 9, (September 2006), pp. 1237-1239, ISSN 0910-6340

Sato, Y. & Sasaki, S. (2008). Observation of Oscillation in Bacterial Luminescence. *Analytical Sciences*, Vol. 24, No. 3, (January 2008), pp. 423-425, ISSN 0910-6340

Sato, Y; Shimizu S; Ohtaki A; Noguchi K; Miyatake H; Dohmae N; Sasaki S; Odaka M & Yohda M. (2010). Crystal structures of the lumazine protein from *Photobacterium kishitanii* in complexes with the authentic chromophore, 6,7-dimethyl- 8-(1'-D-ribityl) lumazine, and its analogues, riboflavin and flavin mononucleotide, at high resolution., *Journal of Bacteriology*, Vol. 192, No. 1, (January 2010), pp. 127-33, ISSN 0021-9193

Septer, AN.; Bose, JL.; Dunn, AK. & Stabb, EV. (2010). FNR-mediated regulation of bioluminescence and anaerobic respiration in the light-organ symbiont Vibrio fischeri. *FEMS Microbiology Letters*, Vol. 306, No. 1, (February 2010), pp. 72-81, ISSN 0378-1097

Shirazy, N. H.; Ranjbar, B.; Hosseinkhani, S.; Khalifeh, K.; Madvar; A. R. & Naderi-Manesh, H. (2007). Critical Role of Glu175 on Stability and Folding of Bacterial Luciferase: Stopped-Flow Fluorescence Study. *Journal of Biochemistry and Molecular Biology*, Vol. 40, No. 4, (July 2007), pp. 453-458, ISSN 1225-8687

Szpilewska, H., A. Czyz and G. Wegrzyn; (2003). Experimental Evidence for the Physiological Role of Bacterial Luciferase in the Protection of Cells Against Oxidative Stress, *Current Microbiology*, Vol. 47, No. 5, (November 2003), pp. 379-382 ISSN 0343-8651

Timmins, G. S., S. K. Jackson and H. M. Swartz; (2001). The Evolution of Bioluminescent Oxygen Consumption as an Ancient Oxygen Detoxification Mechanism, Journal of Molecular Evolution, Vol. 52, No. 4, (April 2001), pp. 321-332, ISSN 0022-2844

Tsuchiya, HM.; Drake, JF.; Jost, JL. & Fredrickson, AG. (1972). Predator-prey interactions of *Dictyostelium discoideum* and *Escherichia coli* in continuous culture. *Journal of Bacteriology*, Vol. 110, No. 3, (June 1972), pp. 1147-1153, ISSN 0021-9193

Tu, S. C.; Lei B.; Liu M.; Tang, C. K. & Jeffers C. (2000). Probing the Mechanisms of the Biological Intermolecular Transfer of Reduced Flavin. *The Journal of Nutrition*, Vol. 130, No. 2, (February 2000), pp. 331-332, ISSN 0022-3166

Urbanczyk, H.; Ast, JC. & Dunlap, PV. (2011). Phylogeny, genomics, and symbiosis of Photobacterium. *FEMS Microbiology Reviews*, Vol. 35, No. 2, (September 2010), pp. 324-342, ISSN 0168-6445

Vaiopoulou, E.; Melidis, P.; Kampragou, E. & Aivasidis, A. (2005). On-line load monitoring of wastewaters with a respirographic microbial sensor. *Biosensors and Bioelectronics* , Vol. 21, No. 2, (December 2004), pp. 365-371, ISSN 0956-5663

Varon, M. & Zeigler, BP. (1978). Bacterial predator-prey interaction at low prey density. *Applied and Environmental Microbiology*, Vol. 36, No. 1, (July 1978), pp. 11-17, ISSN 0099-2240

Wu, BM.; Subbarao, KV. & Qin, QM. (2008). Nonlinear colony extension of Sclerotinia minor and S. sclerotiorum. *Mycologia* , Vol. 100, No. 6, (November 2008), pp. 902-910, ISSN 0027-5514

Yano, Y.; Numata, M.; Hachiya, H.; Ito, S.; Masadome, T.; Ohkubo, S.; Asano, Y. & Imato, T. (2001). Application of a microbial sensor to the quality control of meat freshness. *Talanta* , Vol. 54, No. 2, (April 2001), pp. 255-262, ISSN 0039-9140